T0344913

RADIO RECEIVER TECHNOLOGY

RADIO RECEIVER TECHNOLOGY

PRINCIPLES, ARCHITECTURES AND APPLICATIONS

Ralf Rudersdorfer

In cooperation with

Ulrich Graf
(in I.1, I.2, II.8.1, III.9, IV.5, V.2.3, V.3)

Hans Zahnd
(in I.2.3, I.3, III.6.1, III.9.5)

Translated by Gerhard K. Buesching, E. Eng.

Library of Congress Cataloging-in-Publication Data

Rudersdorfer, Ralf.
    [Funkempfängerkompendium. English]
    Radio receiver technology : principles, architectures, and applications / Ralf Rudersdorfer, Ulrich Graf, Hans Zahnd.
        pages cm
    Translation of: Funkempfängerkompendium.
    Includes bibliographical references and index.
    ISBN 978-1-118-50320-1 (hardback)
    1. Radio–Receivers and reception. I. Graf, Ulrich, 1948- II. Zahnd, Hans. III. Title.
    TK6563.R6813 2013
    621.3841′8–dc23
                                                                                    2013008682

A catalogue record for this book is available from the British Library.

ISBN: 9781118503201

Set in 10/12 Times by Laserwords Private Limited, Chennai, India
Printed and bound in Malaysia by Vivar Printing Sdn Bhd

1 2014

Contents

About the Author

Ralf Rudersdorfer, born in 1979, began his career at the Institute for Applied Physics. He then changed to the Institute for Communications Engineering and RF-Systems (formerly Institute for Communications and Information Engineering) of the Johannes Kepler University Linz, Austria, where he is head of Domain Labs and Technics. His activities included the setting up of a measuring station with attenuated reflection properties/antenna measuring lab and furnishing the electronic labs of the Mechatronics Department with new basic equipment.

He began publishing technical papers at the age of 21. In August 2002 he became a Guest Consultant for laboratory equipment and RF hardware and conducted practical training courses in 'Electronic Circuit Engineering' at the reactivated Institute for Electronics Engineering at the Friedrich Alexander University Erlangen-Nuremberg, Germany. In 2006 he applied for a patent covering the utilization of a specific antenna design for two widely deviating ranges of operating frequencies, which was granted within only 14 months without any prior objections. In the winter semesters 2008 to 2011 the Johannes Kepler University Linz, Austria, commissioned him with the execution of the practical training course on 'Applied Electrical Engineering'.

Rudersdorfer is the author of numerous practice-oriented publications in the fields of radio transmitters and radio receivers, high-frequency technology, and general electronics. Furthermore, he was responsible for the preparation of more than 55 measuring protocols regarding the comprehensive testing of transmitting and receiving equipment of various designs and radio standards issued and published by a trade magazine. During this project alone he defined more than 550 intercept points at receivers. He has repeatedly been invited to present papers at conferences and specialized trade fairs. At the same time he is active in counseling various organizations like external cooperation partners of the university institute, public authorities, companies, associations, and editorial offices on wireless telecommunication, radio technology, antenna technology, and electronic measuring systems.

In the do-it-yourself competition at the VHF Convention Weinheim, Germany, in 2003 he received the Young Talent Special Award in the radio technology section. At the short-wave/VHF/UHF conference conducted in 2006 at the Munich University of Applied Sciences, Germany, he took first place in the measuring technology section. The argumentation for the present work in its original version received the EEEfCOM Innovation Award 2011 as a special recognition of achievements in Electrical and Electronic Engineering for Communication. Already at the age of 17 Ralf Rudersdorfer was active as a licensed radio amateur, which may be regarded as the cornerstone of his present interests.

Owing to his collaboration with industry and typical users of high-end radio receivers and to his work with students, the author is well acquainted with today's technical problems. His clear and illustrative presentation of the subject of radio receivers reflects his vast hands-on experience.

Preface

The wish to receive electromagnetic waves and recover the inherent message content is as old as radio engineering itself. The progress made in technical developments and circuit integration with regard to receiver systems enables us today to solve receiver technology problems with a high degree of flexibility. The increasing digitization, which shifts the analog/digital conversion interface ever closer to the receiving antenna, further enhances the innovative character. Therefore, the time has come to present a survey of professional and semi-professional receiver technologies.

The purpose of this book is to provide the users of radio receivers with the required knowledge of the basic mechanisms and principles of present-day receiver technology. Part I presents realization concepts on the system level (block diagrams) tailored to the needs of the different users. Circuit details are outlined only when required for comprehension. An exception is made for the latest state-of-the-art design, the (fully) digitized radio receiver. It is described in more detail, since today's literature contains little information about its practical realization in a compact form.

The subsequent sections of the book deal with radio receivers as basically two-port devices, showing the fields of application with their typical requirements. Also covered in detail are the areas of radio receiver usage which are continuously developed and perfected with great effort but rarely presented in publications. These are (besides modern radio direction finding and the classical radio services) predominantly sovereign radio surveillance and radio intelligence. At the same time, they represent areas where particularly *sophisticated radio receivers* are used. This is demonstrated by the many examples of terrestrial applications shown in Part II.

A particular challenge in the preparation of the book was the *systematic presentation* of all characteristic details in order to comprehend, understand and evaluate the respective equipment properties and behaviour. Parts III and IV, devoted to this task, for the first time list all receiver parameters in a comprehensive, but easy to grasp form. The description consistently follows the same sequence: Physical effect or explanation of the respective parameter, its acquisition by measuring techniques, and the problems that may occur during measurement. This is followed by comments about its actual practical importance. The measuring techniques described result from experience gained in extensive laboratory work and in practical tests. Entirely new territory in the professional literature is entered in Part IV with the model for an evaluation of practical operation and the related narrow margin of interpretation.

The Appendix contains valuable information on the dimensioning of receiving systems and the mathematical derivation of non-linear effects, as well as on signal mixing and secondary reception. Furthermore, the Concluding Information provides a useful method for converting different level specifications as often encountered in the field of radio receivers.

Easy comprehension and reproducibility in practice were the main objectives in the preparation of the book. Many pictorial presentations were newly conceived, and the equations introduced were supplemented with practical calculations.

In this way the present book was compiled over many years and introduces the reader with a basic knowledge of telecommunication to the complex matter. All technical terms used in the book are thoroughly explained and synonyms given that may be found in the relevant literature. Where specific terms reappear in different sections, a reference is made to the section containing the explanation. Due to the many details outlined in the text the book is well suited as a reference work, even for the specialist. This is reinforced by the *index*, with more than *1,200 entries, freely after the motto*:

When the expert (developer) finds the answer to his story,
spirits rise in the laboratory,
and so one works right through the night
instead of only sleeping tight!

Acknowledgements

The professional and technically sound compilation of a specialized text always requires a broad basis of experience and knowledge and must be approached from various viewpoints. Comments from specialists with many years of practical work in the relevant field were therefore particularly helpful.

My special thanks go to the electrical engineers Harald Wickenhäuser of Rohde&Schwarz Munich, Germany, Hans Zahnd, of the Hans Zahnd engineering consultants in Emmenmatt, Switzerland, and Ulrich Graf, formerly with Thales Electron Devices, Ulm, Germany, for their many contributions, long hours of constructive discussions and readiness to review those parts of the manuscript that deal with their field of expertise. Furthermore, I wish to thank Dr. Markus Pichler, LCM Linz an der Donau, Austria, for his suggestions regarding mathematical expressions and notations which were characterized by his remarkable accuracy and willingness to share his knowledge. Thanks also go to Erwin Schimbäck, LCM Linz an der Donau, Austria, for unraveling the mysteries of sophisticated electronic data processing, and to former Court Counsellor Hans-Otto Modler, previously a member of the Austrian Federal Police Directorate in Vienna, Austria, for proofreading the entire initial German manuscript.

I want to thank the electrical engineer Gerhard K. Büsching, MEDI-translat, Neunkirchen, Germany, for his readiness to agree to many changes and his patience in incorporating these, his acceptance of the transfer of numerous contextual specifics, enabling an efficient collaboration in a cooperative translation on the way to the international edition of this book. My thanks are also due to Dr. John McMinn, TSCTRANS, Bamberg, Germany, for the critical review of the English manuscript from a linguistic point of view.

My particular gratitude shall be expressed to the mentors of my early beginnings: Official Councellor Eng. Alfred Nimmervoll and Professor Dr. Dr. h.c. Dieter Bäuerle, both of the Johannes Kepler University Linz, Austria, as well as to Professor Dr. Eng. Dr. Eng. habil. Robert Weigel of the Friedrich Alexander University Erlangen-Nuremberg, Germany, for their continued support and confidence and their guidance, which helped inspire my motivation and love for (radio) technology.

I wish to especially recognize all those persons in my environment, for whom I could not always find (enough) time during the compilation of the book.

Finally, not forgotten are the various companies, institutes and individuals who provided photographs to further illustrate the book.

May the users of the book derive the expected benefits and successes in their dedicated work. I hope they will make new discoveries and have many 'aha' moments while reading or consulting the book. I want to thank them in advance for possible suggestions, constructive notes and feedback.

Ralf Rudersdorfer
Ennsdorf, autumn 2013

I

Functional Principle of Radio Receivers

I.1 Some History to Start

Around 1888 the physicist Heinrich Hertz experimentally verified the existence of electromagnetic waves and Maxwell's theory. At the time his transmitting system consisted of a spark oscillator serving as a high frequency generator to feed a dipole of metal plates. Hertz could recognize the energy emitted by the dipole in the form of sparks across a short spark gap connected to a circular receiving resonator that was located at some distance. However, this rather simple receiver system could not be used commercially.

I.1.1 Resonance Receivers, Fritters, Coherers, and Square-Law Detectors (Detector Receivers)

The road to commercial applications opened only after the Frenchman Branly was able to detect the received high-frequency signal by means of a coherer, also known as a *fritter*. His *coherer* consisted of a tube filled with iron filings and connected to two electrodes. The transfer resistance of this setup decreased with incoming high-frequency pulses, producing a crackling sound in the earphones. When this occurred the iron filings were rearranged in a low-resistance pattern and thus insensitive to further stimulation. To keep them active and maintain high resistance they needed to be subjected to a shaking movement. This mechanical shaking could be produced by a device called a Wagner hammer or knocker. A receiving system comprising of a dipole antenna, a coherer as a detector, a Wagner hammer with direct voltage source and a telephone handset formed the basis for Marconi to make radio technology successful world-wide in the 1890s.

The components of this receiver system had to be modified to meet the demands of wider transmission ranges and higher reliability. An increase in the range was achieved by replacing the simple resonator or dipole by the Marconi antenna. This featured a high vertical radiator as an isolated structure or an expanded fan- or basket-shaped antenna

Radio Receiver Technology: Principles, Architectures and Applications, First Edition. Ralf Rudersdorfer.
© 2014 Ralf Rudersdorfer. Published 2014 by John Wiley & Sons, Ltd.

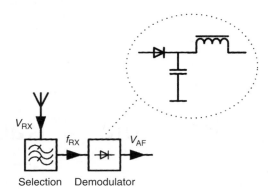

Figure I.1 Functional blocks of the detector receiver. The demodulator circuit shown separately represents the actual detector. With the usually weak signals received the kink in the characteristic curve of the demodulator diode is not very pronounced compared to the signal amplitude. The detector therefore has a nonlinear characteristic. It is also known as a square-law detector. (The choke blocks the remaining RF voltage. In the simplest versions it is omitted entirely.)

of individual wires with a ground connection. The connection to ground as a 'return conductor' had already been used in times of wire-based telegraphy.

The selectivity which, until then, was determined by the resonant length of the antenna, was optimized by oscillating circuits tuned by means of either variable coils or variable capacitors. At the beginning of the last century a discovery was made regarding the rectifying effect that occurs when scanning the surface of certain elements with a metal pin. This kind of detector often used a galena crystal and eventually replaced the coherer. For a long while it became an inherent part of the *detector* receiver used by our great-grandparents (Fig. I.1).

The rapid growth of wireless data transmission resulted in further development of receiving systems. Especially, the increase in number and in density of transmitting stations demanded efficient discriminatory power. This resulted in more sophisticated designs which determined the selectivity not only by low-attenuation matching of the circuitry to the antenna but also by including multi-circuit bandpass filters in the circuits which select the frequency. High circuit quality was achieved by the use of silk-braided wires wound on honeycomb-shaped bodies of suitable size or of rotary capacitors of suitable shape and adequate dielectric strength. This increased not only the selectivity but also the accuracy in frequency tuning for station selection.

I.1.2 Development of the Audion

Particularly in military use and in air and sea traffic, wireless telegraphy spread rapidly. With the invention of the electron tube and its first applications as a rectifier and RF amplifier came the discovery, in 1913, of the feedback principle, another milestone in the development of receiver technology. The use of a triode or multi-grid tube, known as the *audion*, allowed circuit designs that met all major demands for receiver characteristics.

For the first time it was possible to amplify the high-frequency voltage picked up by the antenna several hundred times and to rectify the RF signal simultaneously. The unique feature, however, was the additional use of the feedback principle, which allowed part of the amplified high frequency signal from the anode to be returned in the proper phase to the grid of the same tube. The feedback was made variable and, when adjusted correctly, resulted in a pronounced undamping of the frequency-determining grid circuit. This brought a substantial reduction of the receive bandwidth (Section III.6.1) and with it a considerable improvement of the selectivity. Increasing the feedback until the onset of oscillation offered the possibility of making the keyed RF voltage audible as a beat note. In 1926, when there were approximately one million receivers Germany, the majority of designs featured the audion principle, while others used simple detector circuits.

The nomenclature for audion circuits used 'v', derived from the term 'valve' for an electron tube. Thus, for example, 0-v-0 designates a receiver without RF amplifier and without AF amplifier; 1-v-2 is an audion with one RF amplifier and two AF amplifier stages. Improvements in the selective power and in frequency tuning as well as the introduction of direct-voltage supply or AC power adapters resulted in a vast number of circuit variations for industrially produced receiver models. The general interest in this new technology grew continuously and so did the number of amateur radio enthusiasts who built their devices themselves. All these various receivers had one characteristic in common: They always amplified, selected and demodulated the desired signal at the same frequency. For this reason they were called *tuned radio frequency* (TRF) *receivers* (Fig. I.2).

Due to its simplicity the TRF receiver enabled commercial production at a low price, which resulted in the wide distribution of radio broadcasting as a new medium (probably the best-known German implementation was the 'Volksempfänger' (public radio receiver)). Even self-built receivers were made simple, since the required components were readily available at low cost. However, the tuned radio frequency receiver had inherent technical deficiencies. High input voltages cause distortions with the audion, and circuits with several cascading RF stages of high amplification tend to self-excitation. For reasons of electrical synchronization, multiple-circuit tuning is very demanding with respect to mechanical precision and tuning accuracy, and the selectivity achievable with these circuits depends on the frequency (Fig. I.3). Especially the selectivity issue gave rise to the principle of superheterodyne receivers (superhet in short) from 1920 in the US

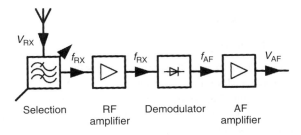

Figure I.2 Design of the tuned radio frequency receiver. Preamplification of the RF signal received has resulted in a linearization of the demodulation process. The amplified signal appears to be rather strong compared to the voltage threshold of the demodulator diode (compare with Figure I.1).

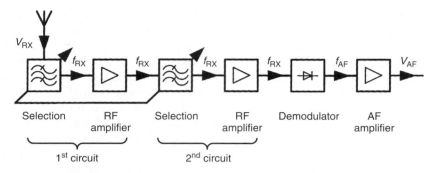

Figure I.3 Multi-tuned radio frequency receiver with synchronized tuning of the RF selectivity circuits. In the literature this circuit design may also be found under the name dual-circuit tuned radio frequency receiver.

and 10 years later in Europe. The superhet receiver solved the problem in the following way. The received signal was preselected, amplified and fed to a mixer, where it was combined with a variable, internally generated oscillator signal (the heterodyne signal). This signal originating from the local oscillator is also known as the LO injection signal. Mixing the two signals (Section V.4.1) produces (by subtraction) the so-called IF signal (intermediate frequency signal). It is a defined constant RF frequency which, at least in the beginning, for practical and RF-technological reasons was distinctly lower than the receiving frequency. By using this low frequency it was possible not only to amplify the converted signal nearly without self-excitation, but also to achieve a narrow bandwidth by using several high quality bandpass filters. After sufficient amplification the intermediate frequency (IF) signal was demodulated. Because of the advantages of the heterodyne principle the problem of synchronizing the tuning oscillator and RF circuits was willingly accepted. The already vast number of transmitter stations brought about increasing awareness of the problem of widely varying receive field strengths (Section III.18). The TRF receiver could cope with the differing signal levels only by using a variable antenna coupling or stage coupling, which made its operation more complicated. By contrast, the utilization of automatic gain control (Section III.14) in the superhet design made it comparatively easy to use.

I.2   Present-Day Concepts

I.2.1   Single-Conversion Superhet

The *superheterodyne receiver* essentially consists of RF amplifier, mixer stage, intermediate frequency amplifier (IF amp), demodulator with AF amplification, and tunable oscillator (Fig. I.4). The high-frequency signal obtained from the receiving antenna is increased in the preamplifier stage in order to ensure that the achieved signal-to-noise ratio does not deteriorate in the subsequent circuitry. In order to process a wide range from weak to strong received signals it is necessary to find a reasonable compromise between the maximum gain and the optimum signal-to-noise ratio (Section III.4.8). Most modern systems can do *without* an RF preamplifier, since they make use of low-loss selection and

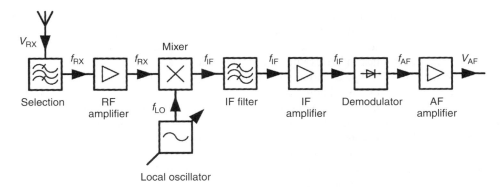

Figure I.4 Functional blocks of the simple superhet. Tuning the receiving frequency is done by varying the frequency of the LO injection signal. Only the part of the converted signal spectrum that passes the passband characteristic (Fig. III.42) of the (high-quality) IF filter is available for further processing.

mixer stages with low conversion loss. The required preselection is achieved by means of a tunable preselector or by using switchable bandpass filters. These are designs with either only a few coils or with a combination of high-pass and low-pass filters.

Previously, the mixer stage (Section V.4) was designed as an additive mixer using a triode tube. This was later replaced by a multiplicative mixer using a multi-grid tube like a hexode (in order to increase the signal stability some circuit designs made use of beam-reflection tubes as mixers). With the continued progress in the development of semiconductors, field-effect transistors were used as additive mixers. These feature a distinct square characteristic and are clearly superior to the earlier semiconductor mixers using bipolar transistors. Later developments led to the use of mixers with metal oxide field-effect transistors (FETs). The electric properties of such FETs with two control electrodes correspond to those of cascade systems and enable improved multiplicative mixing. High oscillator levels result in acceptable large-signal properties (Section III.12). Symmetrical circuit layouts suppressing the interfering signal at the RF or IF gate are still used today in both simple- and dual-balanced circuit designs with junction FETs. Only with the introduction of Schottky diodes for switches did it become possible to produce simple low-noise mixers with little conversion damping in large quantities as modules with defined interface impedances. Measures such as increasing the local oscillator power by a series arrangement of diodes in the respective branch circuit resulted in high-performance mixers with a very wide dynamic range, which are comparatively easy to produce. Today, they are surpassed only by switching mixers using MOSFETs as polarity switches and are controlled either by LO injection signals of very high amplitudes or by signals with extremely steep edges from fast switching drivers [1]. With modern switching mixers it becomes particularly important to terminate all gates with the correct impedance and to process the IF signal at high levels and with low distortion.

The first IF amplifiers used a frequency range between about 300 kHz and 2 MHz. This allowed cascading several amplifier stages without a significant risk of self-excitation, so that the signal voltage suitable for demodulation could be derived even from signals close to the sensitivity limit (Section III.4) of the receiver. Initially, the necessary selection was achieved by means of multi-circuit inductive filters. Later on the application of highly selective quartz resonators was discovered, which soon replaced the LC filters. The use of several quartz bridges in series allowed a bandwidth adapted to the restrictions of the band allocation and the type of modulation used. Since quartz crystals were costly, several bridge components with switchable or variable coupling were used instead. This enabled manual matching of the bandwidth according to the signal density, telegraphy utilization or radiotelephony. Sometime later, optimum operating comfort was obtained by the use of several quartz filters with bandwidths matched to the type of modulation used. Replacing the quartz crystals by ceramic resonators provided an inexpensive alternative. The characteristics of mechanical resonators were also optimized to suit high performance IF filters. Electro-mechanical transducers, multiple mechanical resonators and so-called reverse conversion coils could be integrated into smaller housings, making them fit for use in radio receivers. The high number of filter poles produced with utmost precision were expensive, but their filter properties were unsurpassed by any other analog electro-mechanical system.

Continued progress in the development of small-band quartz filters for near selection (Section III.6) allowed extending the range of intermediate frequencies up to about 45 MHz. Owing to the crystal characteristics, filters with the steepest edges operated at around 5 MHz. Lower frequencies required very large quartz wafers, while higher frequencies affected the slew rate of filters having the same number of poles. Modern receivers already digitize the RF signal at an intermediate frequency, so that it can be processed by means of a high-performance digital signal processor (DSP). The functionality of the processor depends only on the operating software. It not only performs the 'calculation' of the selection, but also the demodulation and other helpful tasks like that of notch-filtering or noise suppression.

The maximum gain, especially of the intermediate frequency amplifier, was adapted to the level of the weakest detectable signal. With strong incoming signals, however, the gain was too high by several orders of magnitude and, without counter measures, resulted in overloading the system. In order to match the amplifier to the level of the useful signal and to compensate for fading fluctuations, the automatic gain control (AGC) was introduced (Section III.14). By rectifying and filtering the IF signal before its demodulation, a direct voltage proportional to the incoming signal level is generated. This voltage was fed to amplifier stages in order to generate a still undistorted signal at the demodulator even from the highest input voltages, causing the lowest overall gain. When the input level decreased the AGC voltage also decreased, causing an increase in the gain until the control function is balanced again. However, the amplifier stages had to be dimensioned so that their gain is controlled by a direct voltage. Very low input signals produce no control voltage, so that the maximum IF gain is achieved. The first superhets for short-wave reception were designed with electron tubes having a noise figure (Section III.4.2) high enough that suitable receiver sensitivities could not be achieved without an RF preamplifier. In order to protect critical mixer stages from overloading, the RF preamplifier was usually integrated into the AGC circuit.

To ensure that signals of low receive field strength and noise were not audible at full intensity, some high-end receivers featured a combination of manual gain control (MGC) and automatic gain control (AGC), the so-called delayed control or delayed AGC (Fig. I.5).

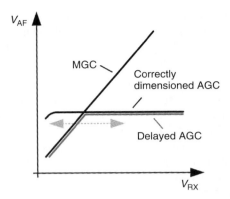

Figure I.5   Functional principle of different RX control methods. In the case of *manual control* the preset gain is kept constant, that is, the AF output voltage follows the RF input voltage proportionally. The characteristic curve can be shifted in parallel by changing the MGC voltage (the required control voltage is supplied from an adjustable constant voltage source). If dimensioned correctly, the automatic gain control (AGC) maintains a constant AF output voltage over a wide range of input voltages. The *delayed AGC* is not effective with weak input signals, but becomes active when the signal exceeds a certain preadjusted threshold and automatically maintains a constant AF output voltage – it is therefore called the 'delayed' gain control.

The automatic control of the gain cuts in only at a certain level, while with lower RF input signals the gain was kept constant. This means that up to an adjustable threshold both the input signal and the output signal increased proportionally. Thus, the audibility of both weak input signals and noise is attenuated to the same degree [2]. This makes the receiver sound clearer. In addition, the sometimes annoying response of the AGC to interfering signals of frequencies close to the receiving frequency (Section III.8.2) that may occur with weak useful signals, can be limited.

During the time when radio signals were transmitted in the form of audible telegraphy or amplitude-modulation signals a simple diode detector was entirely suitable as a demodulator. This was followed by a variable multi-stage AF amplifier for sound reproduction in headphones or loudspeakers. In order to make simple telegraphy signals audible an oscillator signal was fed to the last IF stage in such a way that a beat was generated in the demodulator as a result of this signal and the received signal. When the received signal frequency was in the centre of the IF passband (see Figure III.42) and the frequency of the beat-frequency oscillator deviated by, for example, 1 kHz, a keyed carrier became audible as a pulsating 1 kHz tone. This beat frequency oscillator (BFO) is therefore known as heterodyne oscillator (LO).

With strong input signals the generation of the beat no longer produces satisfactory results. The loose coupling was therefore soon replaced by a separate mixing stage, called the product detector since its output signal is generated by multiplicative mixing. With product detectors it then became possible to demodulate single-side-band (SSB) modulation that could not be processed with an AM detector.

Besides the task of developing a large-signal mixer, a symmetrical quartz filter with steep edges or a satisfactorily functioning AGC (that is well adapted to the modulation type used), especially the design of a variable local oscillator for the superhet presented an enormous challenge for the receiver developer.

The first heterodyning oscillators oscillated freely. Tuning was either capacitive by a rotatable capacitor or inductive after ferrites became available. The first generation of professional equipment used an oscillator resonance circuit that varied synchronously with the input circuits of the RF amplifier stages. For this the variable capacitors had the same number of plate packages as the number of circuits that needed tuning. In most amateur radio equipment, however, the input circuits were tuned separately from the oscillator for practical reasons. Any major detuning of the oscillator therefore required readjusting of the preselector. The frequency of the freely oscillating oscillators was lower than the received frequency. The higher the tuning frequency the lower was the stability with varying supply voltages and temperatures. Frequency stability could be achieved only by utmost mechanical precision in oscillator construction, the integration of cold thermostats, and the use of components having defined temperature coefficients. By combining these measures an optimum compensation was obtained over a wide temperature range. Manufacturing a frequency-stabilized tuning oscillator was difficult, even with industrial production methods, and required extra efforts of testing and measuring.

In order to prevent frequency fluctuations due to changing supply voltages and/or loads, oscillators are usually supplied with voltages from electronically regulated sources. Load variations originating from the mixing stage or subsequent amplifier or keying stages during data transmission are counteracted by incorporating at least one additional buffer stage. Its only task is the electrical isolation of the oscillator from the following circuits.

In the beginning, the receive frequency was indicated as an analog value by means of a dial mounted on the axis of the oscillator tuning element. The dial markings directly indicated the receive frequencies or wavelengths and, in the case of broadcast receivers, showed the stations that could be received. (A few units had a mechanical digital display of the frequency. Among them were the NCX-5 transceiver from National and the 51S-1 professional receiver from Collins. They allowed a tuning accuracy of 1 kHz.)

An accurate reproduction of the tuned-in frequency was possible only with a digital frequency counter used for determining and displaying the operating frequency. The display elements used were Nixie tubes, later the LED dot-matrix or seven-element displays, and recently mostly LC displays. To indicate the receive frequency, the frequency counted at the oscillator must be corrected when resetting the counter either by direct comparison of the BFO frequency counted in a similar manner or by preprogramming the complements.

I.2.2 Multiple-Conversion Superhet

The mixer stage of a superheterodyne receiver satisfies the mathematical condition for generating an intermediate frequency from the heterodyne signal with two different receive frequencies (III.5.3). Both the difference between the receive frequency (f_{RX}) and the LO frequency (f_{LO}) and the difference between the LO frequency and a second receive frequency generate the same intermediate frequency (f_{IF}). The two receive frequencies form a mirror image relative to the frequency of the oscillator, both separated by the IF. The unwanted receive frequency is therefore called the image frequency. The frequency of any such signal is equal to the IF and directly affects the wanted signal or, in extreme cases, covers it altogether. To avoid this, the image frequency must be suppressed. This is usually done by preselection, i.e. by means of the resonance circuits of the RF preamplifier or the preselector. At the beginning of the superhet era the near selection (Section III.6), responsible for the selectivity by filtering the useful signal from the adjacent signals, was possible only with high-quality multi-circuit bandpass filters having a low frequency. From the actual image frequency it is obvious that, for a low IF, it can be suppressed only with a considerable amount of filtering. Especially with receivers designed for several frequency ranges, the reception of high-frequency signals was strongly affected by an insufficiently suppressed image frequency (Section III.5.3). It was therefore necessary to find a compromise between image frequency suppression and selectivity, based on the intermediate frequency.

This problem was solved by twofold heterodyning. To reject the image frequency the first IF was made as high as possible; the higher the IF the lower the effort to suppress the image frequency (see Fig. III.36). A second mixer converted to a second IF so low that good near selection was possible at an acceptable cost (Fig. I.6). But the second mixer again produces both a useful frequency and an image frequency. The second image frequency must also be suppressed as far as possible by means of a filter operating on the first IF. In the era of coil filters this required very careful selection of the frequency.

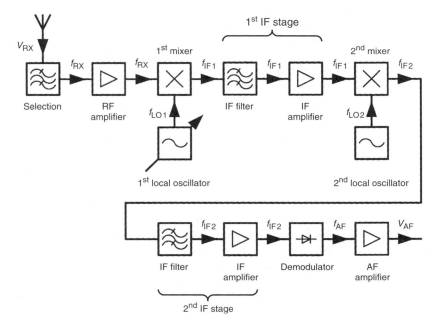

Figure I.6 Operating principle of multiple-conversion superheterodyne receivers. The design shown here is called a dual-conversion superhet. The first IF is a high frequency and serves mainly to prevent receiving image frequencies. The second mixer changes to a lower IF in order to perform the main selection.

The higher the first IF was chosen in the *dual-conversion superhet*, the more difficult it became to manufacture a variable freely-oscillating first local oscillator with a frequency low enough to cause sufficient frequency drifts (Section III.15), for example, for stable telegraphy reception at narrow bandwidths. If the LO frequency was above the receive frequency in one frequency range and below it in the other, the analog frequency scales had to be marked in opposing directions, making operating the equipment cumbersome. Attempts were therefore made to stabilize the first oscillator as well as possible. Initially, this utilized the converter method – the first oscillator remained untuned and was stabilized by a quartz element, while tuning was achieved with the second local oscillator. However, this required that the filter of the first IF be as wide as the entire tuning range. This design was used in almost all early equipment generations for semi-professional use (including amateur radio service) like those produced by Heathkit or Collins. In order to minimize overloading due to the high number of receiving stations within one band, the tuning range was limited to only a few hundred kHz. In the Collins unit, featuring electron tubes, the first IF was merely 200 kHz wide. With a tunable second local oscillator at a lower frequency the conversion to a lower, narrower second IF was simple and stable.

Nevertheless, the problem of large-signal immunity (Section III.12) remained. By using a first tunable local oscillator at a high frequency it was attempted to again reduce the bandwidth of the first IF to the strictly necessary maximum bandwidth, depending on the widest modulation type to be demodulated. At first, the premix system was used. This consisted of a low-frequency tuned oscillator of sufficient frequency stability and a mixer for converting the signal to the required frequency by means of switchable signals from the quartz oscillators. Since the mixing process produced spurious emissions, subsequent filtering with switchable bandwidths was necessary. This is a complex method, but free of the deficiencies described above. It established itself with Drake and TenTec in the semi-professional sector (Fig. I.7). With a tunable first local oscillator it is sufficient for the second LO to use a simple quartz oscillator with a fixed frequency.

As long as the required frequency bands were restricted to a reasonable number (like the short-wave broadcasting bands or the classical five bands of amateur radio services) this principle left nothing to be desired. However, the need for receivers covering all frequency ranges from <1 MHz to 30 MHz inevitably increased the number of expensive quartz elements and increased the demands on near selection of the premixer. This changed only with the availability of low-cost digital integrated semiconductor circuits, which simplified frequency dividing. When dividing the output frequency of an oscillator to a low frequency and comparing it with the divided frequency of a reference signal stabilized by quartz elements, the oscillator can be synchronized by means of a voltage-dependent component (like a varactor diode) using a direct voltage derived from the phase difference between the two signals for retuning the oscillator. This was the beginning of phase-locked loops (PLL) and voltage-controlled oscillators (VCO) (Fig. I.8). Particularly the PLL circuits gave an enormous boost to the advancement of frequency tuning in receivers. Today, highly integrated circuits enable the design of complex and powerful tuned oscillator systems for all frequency ranges. Using several control loops they achieve very high resolution with very small frequency tuning increments [3], short settling times (Section III.15) even with wide frequency variations, and little sideband noise (Section III.7.1). Those circuits used for generating heterodyne signals are called synthesizers.

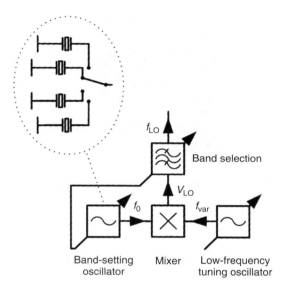

Figure I.7   Architecture of a premixer assembly which feeds an LO injection signal of a stable frequency to the first mixer of a multiple-conversion superhet receiver. The separately depicted circuit design of switchable quartz elements is of course part of an oscillator in actual equipment.

But it is necessary to use processors to make such circuits more ergonomic and the many functions easier to use. With processors the operating frequency can be tuned almost continuously by means of an optical encoder or be activated directly by a number entered via the keyboard. It is possible to store many frequencies in a memory. In the latest developments the loop for fine-tuning is replaced by direct digital synthesis (DDS) (Fig. I.9). This generates an artificial sinusoidal from the digital input information and the signal is tunable in increments of $\ll 1$ Hz. It is controlled by the operating processor,

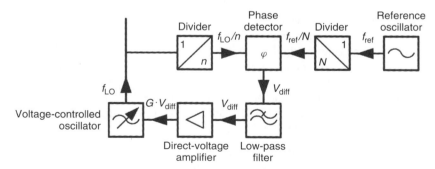

Figure I.8   VCO with phase-locked loop. The direct voltage V_{diff} for automatic frequency tracking is smoothed in the so-called loop filter to prevent spurious signals and sideband noise. V_{diff} is adapted to the required voltage range of the voltage-controlled oscillator via subsequent amplification by the factor G. This results in the constant output frequency $f_{LO} = n \cdot f_{ref}/N$.

Figure I.9 Complete DDS capable of producing output signals up to 400 MHz with a resolution of 14 bits. Only a reference clock and a low-pass filter must be provided externally. (Company photograph of Analog Devices.)

which is required in any case. Depending on the resolution of the D/A converter in the DDS module the output signal generated has very little phase noise (Fig. III.50) and unwanted spurious components (Fig. III.51). Owing to the rapid progress made in this technology DDS generators are currently used in almost every radio receiver. Fully integrated circuits that can generate output signals up to 500 MHz are available. (An example of this technology is AD9912 from Analog Devices, featuring a phase noise as low as -131 dBc/Hz at 10 kHz separation distance with an output frequency of 150 MHz. The output frequency can be varied by increments as small as 3.6 μHz [4]. The spurious emissions actually occurring depend to a large extent on the type of programming.)

It was quickly realized that large-signal problems can be eliminated only if the first narrow-band selection takes place in an early stage of the receive path. In multiple-conversion systems quartz filters with a frequency in the range of about 5 MHz to 130 MHz were therefore included already in the first IF. The first IF is amplified just enough so that the subsequent stages do not noticeably affect the overall noise factor (Section V.1). In high-linearity RF frontends there is no amplification at all upstream of the first mixer. The narrower the bandwidth in the first IF the higher is its relieving effect for the second mixer. Usually the second mixer stage is much simpler than the first mixer. Nowadays, the latest high-end radio receivers match the selected bandwidth already in the first IF stage to the respective transmission method by switching roofing filters (Fig. I.10). (Quartz filters are used in most cases. The commonly used term 'roofing' filter indicates its protective effect on all subsequent stages, just as the roof of a house protects all rooms underneath from the weather.) This satisfies the need for matching the selection to the modulation in order to achieve optimum large-signal immunity or for processing the useful signal with low frequency spacing to strong interferences.

For the second IF, almost all professional receivers used a frequency for which selection filters were readily available on the market, usually the frequency of 455 kHz. Telefunken developed their own mechanical filters of 200 kHz and 500 kHz, while Japanese developers chose to use their own frequencies, probably for competitive reasons. In professional systems amplification was made so high that the AGC cut in even with the weakest signals. This made such signals strong enough to be displayed (Section III.14) and to

Figure I.10 Switchable filters with a bandwidth of 15 kHz/6 kHz/3 kHz matched to the requirements of the transmission methods F3E/A3E/J3E. In modern HF radio transceivers they are placed in the first IF stage (here at a frequency of 64.455 MHz) of the receiving section. Visible are the matching networks arranged close to the actual filters. (Company photograph of ICOM.)

produce a constant AF output level. With these high IF amplifications a control range of 110 dB was no rarity. (For amateur equipment this philosophy never gained ground. Many older-generation radio amateurs were accustomed to the low noise background from their use of low-gain electron tube units which, for weaker signals, needed a 'boost' from the AF amplifier. In order to reproduce such a low background modern amateur receivers also have a low noise level and thus sufficient sensitivity, but the IF amplification is so 'narrow' that only signals with an input voltage of several microvolts produce a signal indication, i.e. a constant output voltage. The control element marked MGC (manual gain control) is often used to shift the threshold value of the delayed AGC (Fig. I.5).)

The demodulation and the AF circuits of a dual-conversion receiver are not much different from those of a single-conversion superhet.

Unlike commercial radio services (Section II.3) that usually work with only a few permanently assigned and sparsely occupied frequencies, search receivers (Section II.4.2) used in radio monitoring, radio reconnaissance and amateur radio services, are dedicated to the reception of weak signals in an interference-prone environment. Very early, those units were therefore equipped with auxiliary devices for interference suppression. Notch filters are used to blank out constant whistling sounds or telegraphy signals from the voice band, while interferences at the periphery of the basic channel can be eliminated by parallel shifting of the filter passband without altering the receiving frequency. The latter method is called passband tuning (Fig. I.11). Eventually, IF systems were developed that allowed independent variation of each of the filter edges (Fig. III.42) of the selection filter in order to respond individually to interferences. A simple passband tuning system can be realized in a single-conversion superhet, while the so-called IF shift for independent edge adjustment always requires a dual-conversion superhet design. When adding the capability to receive signals with frequency modulation (F3E) by means of a dedicated low-frequency limiting IF circuit, all receive functions can be realized in a multiple-conversion superhet receiver. Some units generate a low-frequency IF simply to

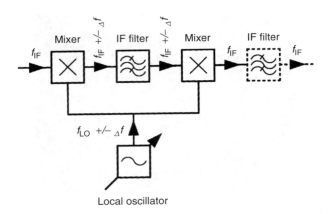

Figure I.11 Passband tuning principle, also known as IF centre frequency shifting. This allows shifting the passband of the IF stage without changing the receive frequency. By including another IF filter behind this stage the IF bandwidth can be varied continuously, that is, it can be matched to the input signal [5]. With this simple method of continuous IF bandwidth adjustment only one filter edge is actually shifted. This makes the passband asymmetrical to the centre frequency. With narrower passbands, however, the shape factor (Section III.6.1) of the IF passband characteristic deteriorates due to the fixed edge steepness of the two IF filters.

enable the use of an efficient notch filter. In order to prevent the mutual interference of the oscillator signals necessary for the multiple-conversion superhet and the resulting mixer products, it is essential not only to plan the frequencies very carefully but also to exercise great care to ensure electronic decoupling and shielding in the mechanical construction.

I.2.3   Direct Mixer

If the oscillator frequency of a superhet receiver is allowed to drift ever closer to the receive frequency the intermediate frequency becomes lower and lower until it reaches zero. The modulation contents of the useful signal are then converted directly to the low frequency range. A receiver working on this principle is called a *direct mixer*, direct-conversion receiver or zero-IF receiver. It avoids the use of an intermediate frequency and thus allows relocating the circuits for amplification, selection and AGC to the AF section (Fig. I.12). This is easily done by using operational amplifiers, such as active filters, amplifiers and control units.

Receivers based on the direct mixer principle remained in the shadows for a long time. The inverse mixing of the signal emitted by the oscillator (Section III.17) with the desired signal leads to hum noise, especially in units operated from a power line. This is why battery-powered units are preferred. With 'simple' heterodyning the image frequency adjacent to the received frequency is also within the baseband. (The baseband is that frequency range that normally contains the useful information, the news contents. In radio technology the transmitted news contents are 'within the baseband' prior to modulation

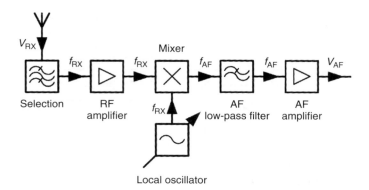

Figure I.12 Basic design of a direct mixer. The image frequency (Section III.5.3) is identical with the incoming frequency f_{RX}. Image frequency reception provides the same tuned frequency, but the demodulated signal spectrum appears inverted, indicating an interference signal.

and after demodulation.) Directly adjacent signals at the image frequency can, therefore, not be suppressed. For a long time this was regarded as such a serious disadvantage that there appeared to be no promise of developing this design to a high-performance 'station receiver'. But systematic implementation of RF/AF engineering enables the direct mixer to provide good receiving performance.

By in-phase splitting of the received signal behind the RF preamplifier and by feeding the two resulting signals to two mixers, where they are converted with the same oscillator signal into two basebands, the two basebands are vectorially orthogonal as AC voltage indicators, provided that the split oscillator signal is also fed to one of the two 90° out of phase (Fig. I.13). *Using these two orthogonal basebands allows the demodulation of signals of all modulation types!* One baseband represents the real component and the other the imaginary component of the complex signal (see also Section I.3.3). Other commonly used terms for these so-called quadrature signals are:

- For the real component: I component or in-phase component.
- For the imaginary component: Q component or quadrature-phase component.

Owing to the fact that RF amplifiers, mixer stages and both baseband branches can be integrated and that after digitization the baseband signals can not only be selected but also demodulated by a highly integrated digital signal processor (DSP), this principle was soon adopted for use in GSM technology. Today, it forms the basis of RF receivers in almost any mobile phone. In mobile radio technology (Section II.3.5) the system is usually referred to as a *homodyne receiver*. Another name for this version of a direct mixer is *quadrature receiver*. Due to the lack of synchronization between the received frequency and the frequency of the LO injection signal, a frequency error occurs because of the limited accuracy even when tuning to nominally the same frequency. For proper functioning [2] this error must be kept small compared with the receive bandwidth (Section III.6.1), since

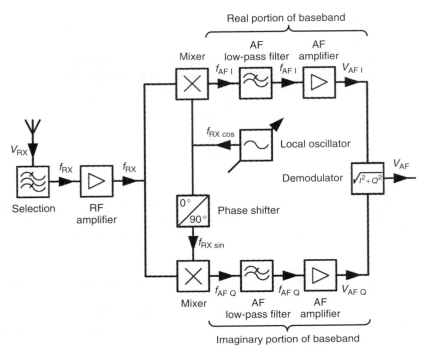

Figure I.13 With the quadrature receiver the main selection is achieved by AF low-pass filters in the I path and the Q path. High performance data can be achieved with fully digitized receiver designs (Section I.2.4) thanks to the very accurate signal processing which this principle makes possible.

slight deviations do not cause any interference, as can be demonstrated mathematically for AM reception:

$$S(t) = \sqrt{(A(t) \cdot \sin(\omega \cdot t))^2 + (A(t) \cdot \cos(\omega \cdot t))^2}$$

$$= A(t) \cdot \sqrt{\sin^2(\omega \cdot t) + \cos^2(\omega \cdot t)}$$

$$= A(t) \tag{I.1}$$

where
$S(t)$ = demodulated AF signal at time (t), in V
$A(t)$ = AM signal at time (t), in V
$\omega \cdot t$ = difference between carrier frequency and LO frequency, in rad
t = considered time, in sec

The term ω is not contained in the result, which proves that the frequency deviation from the LO injection signal is insignificant. This presumes, however, that the two mixed spectra are symmetrical to the LO frequency. This is not the case with selective fading. In this respect this demodulator is inferior to the synchronous receiver. For SSB reception the quadrature receiver requires another 90° phase shifter to enable suppressing the

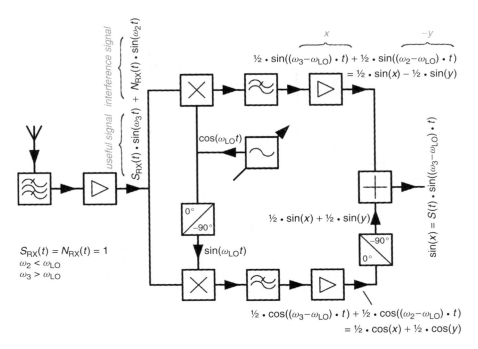

Figure I.14 The quadrature receiver with sideband suppression requires an additional 90° phase shifter. With the fully digitized unit (Fig. I.24) a sideband suppression of more than 100 dB can be obtained without problems.

unwanted sideband (Fig. I.14). The constant phase shift over several octaves in the AF range presented a major challenge in analog technology. This may be another reason why this type of receiver was rarely seen in earlier times.

Synchronizing the LO injection signal with the receive frequency by means of a phase control loop, can accomplish demodulation of FM/PM and AM signals without a demodulator. Such a design is called a *synchronous receiver* (Fig. I.15) which, apart from the omission of the demodulator, is identical to the quadrature receiver. Because of the strictly identical carrier frequencies of the signal and image behind the mixing stage, the even AM sidebands are the same in phase and shape. The same is true for the uneven FM/PM sidebands, assuming 90° out-of-phase mixing in the second branch. In each case, the other component is canceled out. Thus, demodulation takes place during the mixing process [2].

I.2.4 Digital Receiver

All functional blocks of the receiver designs discussed so far can be described mathematically (with regard to the time domain and frequency domain of the transfer characteristics). This means that basically all stages can be reproduced by algorithms in a fast digital processor, provided that A/D conversion (Section I.3.2) is sufficiently fast to convert the signals to a form (bit sequence) suitable for processing. The same considerations

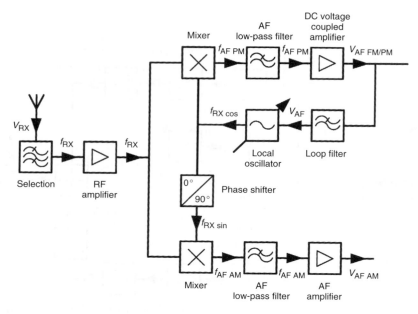

Figure I.15   The synchronous receiver is the second design of the direct mixer that receives the signal without image. If the signal received is phase modulated with a modulation frequency above the limit frequency of the PLL loop filter, the modulation contents can be extracted from the upper branch. Demodulated AM signals are available at the end of the lower branch.

apply as for conventional circuit designs. The in-principle ideal digital architecture has its deficiencies in quantization effects.

In the units marketed from around 1980, digital components were used only for control functions and audio signal processing. These were *first generation digital receivers.*

Using digital signal processors at low intermediate frequencies for 'computationally' processing the useful signal received has been standard in high-end equipment for several years. Modern receivers select the desired signal by means of a DSP from the signal spectrum of the input bandpass converted by the mixer to the intermediate frequency. The DSP performs arithmetic demodulation and keeps the useful signal free of interferences like continuous carrier whistling, noise or crackle. It then evaluates the signal and provides an AGC criterion for controlling the overall gain [6]. A modern DSP is capable of performing the required computing in 'real-time', i.e. with a time delay that is no longer subjectively detectable. Today, these units are called *second generation digital receivers* (Figs. I.16 and I.17). Despite arithmetic processing of the signal, considered unusual from analog perspectives, the significant advantages of this technology are cost savings, particularly for expensive quartz filters, and the enormous flexibility of the characteristics as a result of the software. The analog components, including the RF frontend, must meet high RF demands since these essentially determine the overall receiver properties (III). However, well-functioning digital signal processing alone is by no means sufficient for the manufacture of a radio receiver suitable for practical applications.

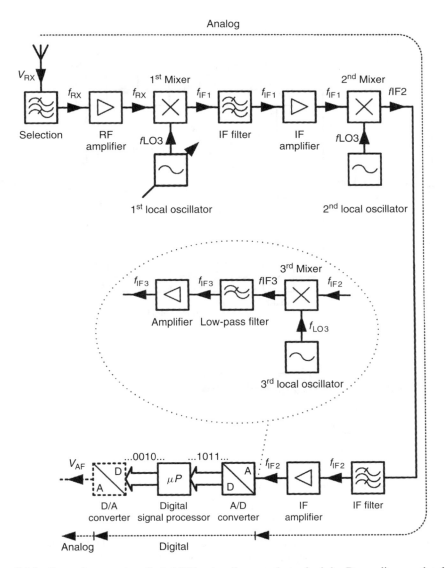

Figure I.16 Second generation digital RX using the superhet principle. Depending on the design concept, A/D conversion is achieved either by subsampling (Section I.3.7) or by the circuitry inside the dotted oval. The fast IF filter, having a bandwidth equal to the widest signal type to be demodulated, guarantees a limitation of the signal frequencies reaching the A/D converter, thus preventing phantom signals (such as those caused by aliasing).

The circuit shown separately depicts the components used for the additional conversion to a lower 3rd IF (usually with a frequency between 12 kHz and 48 kHz). The A/D conversion takes place behind the low-pass filter, having a limit frequency slightly below half the sampling rate. The signal has then passed three mixers and some filters (often too wide for narrow emission classes). (This principle is used in many radio receivers for semi-professional use, as well as in equipment like the VLF/HF receiver EK896 from Rohde&Schwarz.)

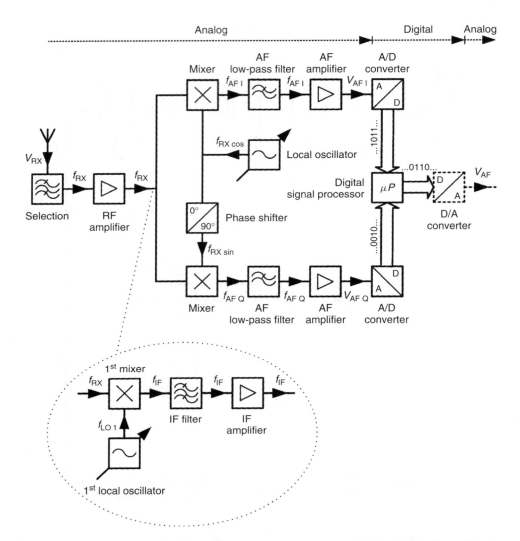

Figure I.17   Second generation digital RX using the quadrature principle. Direct mixers (Section I.2.3) of this design perform a separate A/D conversion of the basebands (as well as of the real and imaginary signal components), which are then combined for subsequent demodulation.

In a different version, shown as a separate circuit, the receive spectrum is converted to a first IF in a highly linear mixer and is then selected by a narrow-band IF filter. This frees the subsequent IQ mixer from sum signals. (The principle was used in the mid 1990s in model 95S-1A from Rockwell-Collins. It covers a receive frequency range from 500 kHz to 2 GHz.)

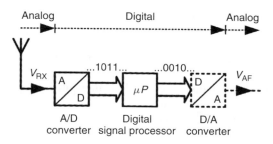

Figure I.18   In an ideal all-digital receiver the A/D conversion takes place close to the antenna socket. The entire signal processing is done by the DSP using mathematical algorithms. However, due to the limited sampling speed of A/D converters, at least one additional low-pass filter is required between the antenna and the A/D converter to prevent exceeding the Nyquist frequency and to avoid aliasing.

Almost all well-known manufacturers of radio equipment [7] now make use of this advanced technology. (The diagram in Figure I.20 shows the classification of the various digital receiver designs according to the location of the A/D converter within the receiver layout.)

In recent years, this technology has made significant progress in digital resolution and clock speed. It seems reasonable therefore to design receivers using only digital signal processing (Fig. I.18). After band selection and analog RF amplification, which is still necessary to achieve a sufficient signal-to-noise ratio, the RF signal is fed directly to a fast A/D converter with high signal dynamics. The subsequent digital signal processor performs all functions previously executed in analog mode, like amplification, selection, interference elimination, and demodulation. The processed signal can now be subjected to digital/analog (D/A) conversion, so that only the resulting AF signal has to be amplified for feeding, for example, a headset or loudspeaker (Fig. I.19). For further signal processing

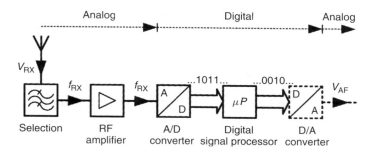

Figure I.19   Digital receiver of generation 2.5. The first units of this type are currently available covering a receive frequency range up to approximately 50 MHz. Depending on the required quality level, it is possible to produce models using only a low-pass filter behind the antenna input instead of a circuit for the specific selection of the desired receiving band.
The final D/A converter is of importance only if the demodulated signal must be available in analog mode, for example, for loudspeakers.

in digital mode, like in decoders or for screen displays, the last D/A converter stage is no longer necessary. Such receiver designs offer a number of advantages [8]:

- Digital signal processing is free of any distortion. Only initial signal conditioning requires special care.
- Problems experienced in analog circuits, like unwanted coupling effects, whistling sounds, and oscillating tendencies, do not exist.
- All modulation modes from AM to complex modes, like quadrature amplitude modulation (QAM) or code-division multiple access (CDMA, Section II.4.1), are supported by one and the same hardware. By using suitable software it is possible to design a multitude of receiver versions up to multi-standard platform models.
- New functions, extensions and modifications of radio standards, like conceptual improvements, can be added by simply installing an improved operating software version (firmware).
- Hardware expenditures based on the effective component costs are much lower than those for analog versions.
- The accuracy is scalable. With suitable software the display of, for example, the relative receive signal strength (Section III.18) can reach an accuracy of better than ± 1 dB over a range of 120 dB.
- Reproducibility is unrestricted. A filter trimmed to a certain shape factor (Section III.6.1) has exactly the same properties in every unit.
- Filter characteristics are freely definable over a wide range of values. This was also desirable with analog filters, but for physical reasons, could not be achieved.

However, with these concepts the technical data of high-end analog receivers can only be partly achieved despite the realization of some still extremely costly professional solutions and first interesting research results (Section I.3) as well as a few experimental models produced by the amateur radio services. For professional use there are already some solutions, however these are still very costly. But owing to new and continuously improved components the feasible range of receiving frequencies is being constantly extended to higher frequencies.

Presently, especially the interference-free dynamic range (Section I.3.2) of A/D converters is still inferior to that of high-performance mixers in combination with narrow-band analog signal processing. The demands on A/D converters regarding a wider bandwidth and a larger dynamic range (to do away with extensive analog prefiltering) are diametrically opposed to each other [7]. It is almost impossible to achieve both goals simultaneously. The best performance is therefore obtained with hybrid concepts (Figs. I.16 and I.17), using analog circuits to generate the IF and digital processing after the respective preselection by quartz filters.

I.2.4.1   Software Radio and Software-Defined Radio

Professional terminology sometimes differentiates between software radio and software-defined radio (SDR) [9]. The first term refers to the ideal *software radio*, i.e. a fully digitized receiver (Fig. I.18). (As already indicated, the software runs on generally available hardware. Since it is primarily the software which defines the functionality of the unit, this is also known as the ideal software radio.)

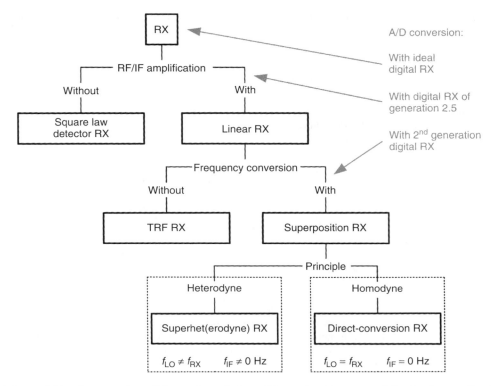

Figure I.20   Survey of possible receiver designs. The various concepts differ fundamentally in their complexity and achievable properties (Part III).

The collective term *software-defined radio* includes all solutions having deficiencies in one or several aspects but which pursue the basic ideas and advantages of a software radio, while considering its technical and economic feasibility on the basis of the hardware available [10]. A/D conversion takes place as close to the antenna as possible (Fig. I.20).

Table I.1 reviews the advantages and disadvantages of current receivers.

I.3   Practical Example of an (All-)Digital Radio Receiver

Already in 1988 reference was made to the technology of fully digitized receivers [11]. The prognosis was made that 'despite all optimism digital receivers of satisfactory quality will hardly appear on the market before the middle of the 1990s ...'. In fact it took even longer, since really usable chipsets have only been developed in the laboratories of various renowned semi-conductor manufacturers within the last few years. The concept (Section I.2.4) of an all-digital receiver (ADR) outlined above will now be described in more detail. State-of-the-art high-quality professional receivers covering a frequency range up to 30 MHz are represented by the first commercially available units, like the ADT-200A from the Swiss engineering consultants Hans Zahnd (Figs. I.21 and I.22) or the MSN-8100-H from Thales Communications, developed for tactical marine communication (both

Table I.1 Principle-related advantages and disadvantages of today's receiver concepts according to [12]

Advantages	Disadvantages
Single-conversion superhet	
+ Most common receiver architecture	− Image signals inherent to the operating principle
+ Good selectivity	− Spurious signal reception
+ Least distortion and highest dynamics achievable with single heterodyning	− IF filter usually not integratable
Low-IF superhet	
+ Can be integrated in monolith	− Image frequency rejection is very sensitive to tolerances
	− Emission of LO injection signal
Multiple-conversion superhet	
+ Excellent receiving characteristics	− Very demanding in design, energy consumption, and number of components
+ Best receiver concept, since a partly digital system	− IF filter can usually not be integrated
Direct mixer	
+ Requires relatively few components	− Emission of LO injection signal
+ Can be entirely integrated in monolith	− Limitations due to inherent parasitic coupling and non-ideal components
+ Potentially low energy consumption	− Very low dynamics
Digital receiver >2nd generation	
+ Can also be integrated in monolith	− Very high energy consumption
+ Flexibly adaptable to changing receiver requirements	− Requires extremely fast linear A/D conversion
	− Requires significant computing power for receiving algorithm
	− Limited dynamics

systems contain an additional digital transmit path, so that the complete unit may correctly be called an all-digital transceiver (ADT)). Especially tailored to the needs of modern radio monitoring (Section II.4) is the EM510 model of Rohde&Schwarz. Controlling the DRM emissions (Section II.6) in full conformity is a feature of model DT700 from the Fraunhofer Institute for Integrated Circuits (Fig. II.47). In these units, an A/D converter samples the sum signal over the receiving range using a sample frequency of more than double the highest possibly frequency received and forwards the information as a parallel bitstream to the signal processing circuitry. Signal processing is all digital and software-controlled. For this the architecture of the homodyne receiver (Section I.2.3) is particularly advantageous [2], since it performs the main selection with comparatively few arithmetic operations. Equipment of this type is casually dubbed *direct receiver* by analogy to direct mixer receivers with their conventional circuitry and because they sample the RF receiving band directly without any conversion.

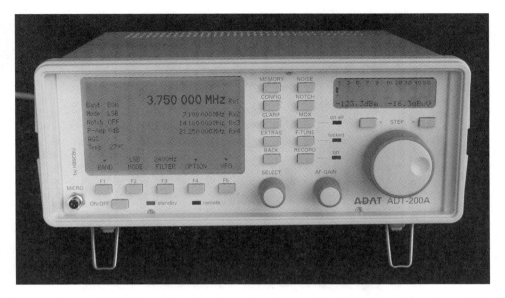

Figure I.21   ADT-200A is a first prototype of an almost fully digitized radio receiver (Digital RX of generation 2.5), designed as a stand-alone unit. Using a 14 bit A/D converter with a signal-to-noise ratio of 74 dB above half the Nyquist bandwidth of 36.86 MHz, in combination with the subsequent decimation it achieves a dynamic range (Section III.9.7) obtained so far only in high-end multiple-conversion superhets. A high-performance signal processor of the latest generation from Analog Devices having a processing power of up to 2 billion instructions per second performs the actual signal processing. (Company photograph of Hans Zahnd engineering consultants.)

I.3.1   Functional Blocks for Digital Signal Processing

The entire frequency range from DC to 30 MHz is fed to an A/D converter via a steep low-pass filter with a limit frequency of 30 MHz (Fig. I.19). The task of the low-pass filter is to prevent frequencies above half the sampling rate (32.5 MHz in this example) from reaching the A/D converter. The A/D converter is the link between analog and digital signal processing. This block essentially determines the receiver properties and should therefore be given special care! To achieve the high performance of a multiple-conversion receiver (like, for example, those of the 95S-1A from Rockwell-Collins or the TMR6100 from Thales Communications) in analog design up to the last IF stage would require a converter of at least 17 bits. The first 12 bit converters having a suitable speed became available in the year 2000. With the ADS852, Burr Brown was one of the first manufacturers to offer a 12 bit converter with a sampling rate of 65 mega-samples per second (MS/s) that was of high quality and still affordable [13]. With the AD6645, Analog Devices offered an improved component featuring 80 MS/s or 105 MS/s [14] and also includes a 16 bit converter, the AD9446 [15], in its sales program. The Linear Technology model LTC2208 is available in versions with 14 bit or 16 bit resolution and 130 MS/s [16]. What are the receiver characteristics that can be expected from such components?

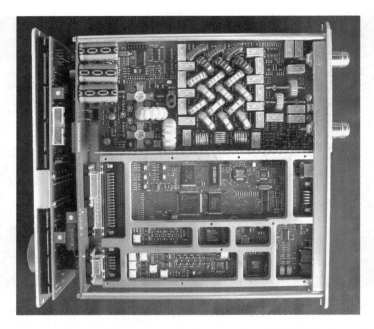

Figure I.22 Hardware layout of the communications receiver ADT-200A shown in Figure I.21. The lower section illustrates the module performing the digital signal processing (Fig. I.29). The upper section shows the 50 W HF transmit output stage. (Company photograph of Hans Zahnd engineering consultants.)

I.3.2   The A/D Converter as a Key Component

An ideal A/D converter is capable of splitting the input signal into 2^n equal voltage components, thus a 14 bit converter with an input voltage range of, for example, 1 V can convert this range into 2^{14} portions of 61 µV each. Any values in between are rounded off. Rounding errors cause noise, the so-called quantization noise. The theoretically possible signal-to-noise ratio (Section III.4.8) for sinusoidal signals is

$$SNR = Bit_{spec} \cdot 6.02 \text{ dB} + 1.76 \text{ dB} \tag{I.2}$$

where
SNR = signal-to-noise ratio of an ideal A/D converter, in dB
Bit_{spec} = specified resolution of the A/D converter, in bits

According to this equation the signal-to-noise ratio of the 14 bit converter considered would be

$$SNR = 14 \text{ bit} \cdot 6.02 \text{ dB} + 1.76 \text{ dB} = 86 \text{ dB}$$

In reality it is not possible to approach this ideal value. The data sheet for this converter specifies $SNR = 75$ dB. This first impression is not very encouraging, since the noise level of 75 dB below 1 V corresponds to a voltage of 178 µV or S9 + 11 dB (Section III.18.1).

However, this noise level is in relation to the entire bandwidth of 32.5 MHz (the Nyquist bandwidth). The A/D converter generates an enormous bitstream of

$$14 \text{ bit} \cdot 65 \text{ MS/s} = 0.91 \text{ GBit/s}$$

This includes the entire receive signal contents, from DC to 30 MHz! In fact, however, only a very narrow portion of it, the receive bandwidth (Section III.6.1), is of interest for, for example, the demodulation of an SSB signal. The huge amount of data needs to be reduced. In signal processing the reduction of the sampling rate is called decimation (see Fig. I.24) and is performed by a special digital filter [17] that averages the signal values of a certain number of sample values and forwards them with a reduced number of (combined) sample values to the subsequent decimation stage. Averaging reduces the quantization noise, resulting in a process gain:

$$G_{\text{dB p}} = 10 \cdot \lg \left(\frac{f_{\text{s}}}{2 \times B_{-6\,\text{dB}}} \right) \qquad (\text{I.3})$$

where
$\quad G_{\text{dB p}}$ = process gain figure by decimation, in dB
$\qquad f_{\text{s}}$ = sampling rate of the A/D converter, in S/s
$\quad B_{-6\,\text{dB}}$ = receive bandwidth (-6 dB bandwidth) of the receive path, in Hz

With a receive bandwidth of 2.4 kHz, as is common for demodulating class J3E emissions, and a sampling rate of 65 MS/s the resulting process gain figure is

$$G_{\text{dB p}} = 10 \cdot \lg \left(\frac{65 \text{ MS/s}}{2 \cdot 2.4 \text{ kHz}} \right) = 41.3 \text{ dB}$$

This reduces the initially calculated noise floor from 178 μV to 1.53 μV. To achieve the usual value of the input noise voltage (Section III.4.7) of 0.2 μV EMF, it is necessary to include an upstream preamplifier. In fact, high intermodulation immunity (Section III.9.6) is possible only without or at most with low RF preamplification. This remains the weakest point of this concept.

The 14 bit A/D converter LTC2208 features a third-order intercept point (Section III.9.8) as high as 47 dBm, which can only be achieved with a high input noise level (Section III.4.2) of almost 30 dB noise figure (Fig. I.23). To obtain a receiver noise figure of $F_{\text{dB}} = 12$ dB by using a preamplifier ($F_{\text{dB}} = 6$ dB) would require a high amplification of 19 dB. Despite this high amplification a total intercept point ($IP3_{\text{tot}}$) of more than 25 dBm is possible, provided that the preamplifier alone has an output $IP3$ of more than 50 dBm. These high demands can be somewhat relaxed if the A/D converter is accessed by an impedance transformer. The input impedances of the various blocks are between 100 Ω and 800 Ω. The amplification can be reduced by 12 dB, which under favourable conditions may result in an $IP3$ of more than 30 dBm. However, the intermodulation response of an A/D converter cannot be compared to that of analog non-linear circuits. Section III.9.5 gives a more detailed description.

The theoretically achievable properties of various A/D converters as determined by the methods of calculation described are shown in Table I.2. The values stated have been confirmed to a large extent by practical tests.

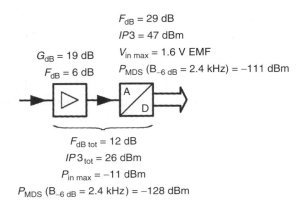

$$F_{dB} = 29 \text{ dB}$$
$$IP3 = 47 \text{ dBm}$$
$$G_{dB} = 19 \text{ dB} \qquad V_{in\ max} = 1.6 \text{ V EMF}$$
$$F_{dB} = 6 \text{ dB} \qquad P_{MDS} (B_{-6\ dB} = 2.4 \text{ kHz}) = -111 \text{ dBm}$$

$$F_{dB\ tot} = 12 \text{ dB}$$
$$IP3_{tot} = 26 \text{ dBm}$$
$$P_{in\ max} = -11 \text{ dBm}$$
$$P_{MDS} (B_{-6\ dB} = 2.4 \text{ kHz}) = -128 \text{ dBm}$$

Figure I.23 Characteristic properties of the components for the calculation described and the resulting overall parameters.

Owing to the relatively high amplification the maximum signal level at the receiver input is reduced to the problematic value of less than -11 dBm. For LW/MW/SW reception even higher (sum) receive levels may be available from high-performance antennas [18]. The use of an attenuator from 0 to 25 dB controlled by the AGC (Section III.14) is an effective counter-measure. This shifts the dynamic range (Section III.9.7), which causes a response to the different signal strengths of the receiving bands. (Such a measure for signal conditioning has already been described in [19]. It stipulates that a switchable level attenuator be inserted before the analog receive section (the RF frontend) and a control amplifier be introduced between the analog receive section and the converter. The concept was initially designed for a second-generation digital receiver. The attenuator is controlled at the input of the control amplifier by the signal level. This arrangement is said to provide optimum utilization of the dynamic range of the A/D converter and a high immunity against overloading.) In the unit shown in Figure I.21 the automatic inclusion of such an attenuator complies with the following principle:

- The sum signal from the A/D converter is monitored at the input of the digital down converter (Fig. I.24). If the peak value exceeds the value 1 dB below the overload point several times within 1 second, the warning 'Intermodulation!' is displayed.
- One second later, the attenuation is incremented by 5 dB.
- If the sum signal remains 8 dB over the overload point for at least 5 seconds the attenuator switches back one increment (5 dB) until the originally preset value is restored.

Since the indication of the relative strength of the receive signal compensates the values of the attenuator and preamplifier, the process remains unnoticed by the operator. The sensitivity (Section III.4) of the receiver of course decreases with an increase in the attenuator damping, but in most cases remains below the external noise received [18].

A significant improvement of the large-signal properties (Section III.12) can be obtained by means of sub-octave filters, such as used in conventional receivers. Such filters reduce

Table I.2 Calculated parameters (Part III) of digital receivers using the components described

	AD6645	AD9446	LTC2288
Sensitivity			
Max. input voltage ($=0\,$dBc)	$2.2\,\mathrm{V_{pp}}$	$3.2\,\mathrm{V_{pp}}$	$2.25\,\mathrm{V_{pp}}$
Recommended source impedance (Z)	$800\,\Omega$	$800\,\Omega$	$100\,\Omega$
Max. input level ($P_{\mathrm{in\,max}}$) with matched Z	$-1.2\,$dBm	$2.0\,$dBm	$8.0\,$dBm
Signal-to-noise ratio***	$75\,$dB	$81\,$dB	$78\,$dB
Noise level in 1st Nyquist band	$-76.2\,$dBm	$-79\,$dBm	$-70\,$dBm
Process gain figure ($G_{\mathrm{dB\,p}}$)	$41.3\,$dB	$41.3\,$dB	$41.3\,$dB
Minimum discernible signal ($P_{\mathrm{MDS\,A/D}}$) at A/D converter	$-117.5\,$dBm	$-120.3\,$dBm	$-111.3\,$dBm
Noise figure ($F_{\mathrm{dB\,A/D}}$) of the A/D converter	$22.7\,$dB	$19.9\,$dB	$28.9\,$dB
Preamplification figure ($G_{\mathrm{dB\,preamp}}$) for $F_{\mathrm{dB\,tot}}=12\,$dB**	$11.9\,$dB	$9.1\,$dB	$18\,$dB
Minimum discernible signal (P_{MDS}) of overall RX	$-128\,$dBm	$-128\,$dBm	$-128\,$dBm
Operational sensitivity at $50\,\Omega$, $10\,$dB $(S+N)/N$	$0.28\,\mu$V	$0.28\,\mu$V	$0.28\,\mu$V
Dynamic range of preamplifier at a receive bandwidth $B_{-6\,\mathrm{dB}}=2.4\,$kHz	$114.9\,$dB	$120.9\,$dB	$118\,$dB
Third-order intercept point (IP3)			
Intermodulation ratio ($IMR3_{\mathrm{A/D}}$) of the A/D converter at $-7\,$dBc***	$-90\,$dBc	$-96\,$dBc	$-93\,$dBc
$IP3_{\mathrm{A/D}}$ of the A/D converter*	$36.8\,$dBm	$43.0\,$dBm	$47.5\,$dBm
Required output $IP3_{\mathrm{preamp}}$ of the preamplifier	$45\,$dBm	$45\,$dBm	$50\,$dBm
$IP3$ of the overall RX	$24.3\,$dBm	$31.8\,$dBm	$24.1\,$dBm
Maximum intermodulation-limited dynamic range (ILDR) of the overall RX			
$ILDR = 2/3 \cdot (IP3 - P_{\mathrm{MDS}})$	$101.6\,$dB	$106.6\,$dB	$101.5\,$dB

*$IP3_{\mathrm{A/D}} = (P_{\mathrm{in\,max}} - 7\,\mathrm{dB}) + IMR3_{\mathrm{A/D}}/2$.

**$G_{\mathrm{preamp}} = (F_{\mathrm{A/D}} - 1)/(F_{\mathrm{tot}} - F_{\mathrm{V}})$; $F_{\mathrm{dB\,V}} = 6\,$dB; (F: numerical value, not in dB).

***according to datasheet [14], [15], [16].

Table I.2 contains the theoretically achievable properties of a digital RX of generation 2.5 with commercially available A/D converters with 65 MS/s sample rate. Not taken into consideration are the attenuation through upstream filters and the influence of sideband noise (Section III.7.1) of the reference clock oscillator on the noise factor of the converter. It is particularly important that the preamplifier is correctly matched to the converter input, which is usually of high resistance. Since the manufacturer of the LTC2208 recommends a source impedance of $100\,\Omega$, this will perform rather poorly.

the process dynamics, since the difference between the weakest and the strongest signals (or the sum voltage) within the passband of the respective bandpass filter is lower than the entire short-wave range.

An important and, at the same time, critical parameter of any A/D converter is the spurious-free dynamic range (SFDR). This defines the ratio of the (unwanted) signal mix of higher order to the maximum input signal. These mixing products are caused by interference between the input signal and the sampling frequency f_{s}, whereby products

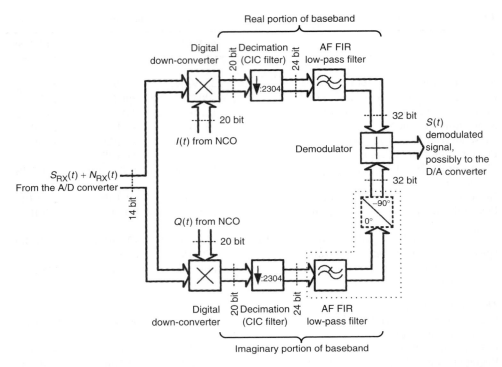

Figure I.24 Design of IF zero mixing with demodulation in a fully digitized mode. Decimation for reducing the sampling rate, without which meaningful post-processing by means of DSP would not be possible, is done by a CIC (cascaded integrator comb) filter. SSB demodulation requires an additional phase shifter in the Q path.

larger than $f_s/2$ are the result of so-called aliasing in the range between DC and $f_s/2$ (see also Fig. I.32). The datasheets for modern 14 bit A/D converters specify an SFDR of more than 100 dB.

Even better results can be achieved by using specific sigma delta A/D converters [9], which perform noise shaping at a sufficiently high oversampling rate [20]. The higher the order of the sigma delta A/D converter the more quantization noise is shifted out of the desired frequency range without the need to increase oversampling. Their use in digital receivers has been investigated for some time [21, 22].

I.3.3 Conversion to Zero Frequency

Selecting the desired signal is similar to selecting with conventional receivers, that is by mixing (Section V.4) with the signal from a local oscillator, utilizing the principle of direct-conversion receivers. The local oscillator is tuned exactly to the carrier of the signal to be received. This principle corresponds to that of a synchronous receiver (Fig. I.15) for

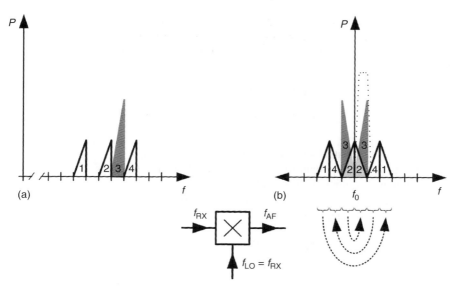

Figure I.25   Overlapping with direct mixer caused by mixing at zero position. Graph a) shows a possible signal scenario in the RF frequency band, and graph b) shows the same after shifting to zero position by mixing for the subsequent demodulation. Interference signals 1 and 2 appear mirrored about f_0. This presents no problem for AM, since both sidebands are symmetrical to the carrier and have identical information contents. However, with SSB this means that interference signal 2, directly adjacent to the desired signal 3, is displayed in the reverse position.

AM, mixing both sidebands of the AM signal in the audio frequency region. The carrier is therefore exactly at zero frequency. This causes the lower sideband of the AM signal to be in a negative frequency range. It can be demonstrated mathematically that the signal on the negative frequency axis is folded around the zero frequency to the positive frequency axis (Fig. I.25). For AM this presents no problem, since the two sidebands are symmetric to the carrier and have identical information contents. With SSB this situation is different; this process causes a reversal of the interference signal immediately adjacent to the useful signal in the basic channel. This problem must also be solved with the conventional analog direct-conversion receivers. The DC receiver uses two mixers controlled by a quadrature LO signal of $0°$ and $90°$ phases, as described in Section I.2.3. This produces a real component and an imaginary component (also called I channel and Q channel), which are fed separately to a low-pass filter (Fig. I.13). In order to suppress the unwanted sideband, it is necessary to shift the phase of the Q path again by $-90°$ after demodulating the SSB (Fig. I.14).

In SSB modulators this method is also known as the phase method. This principle is easily explained mathematically on the basis of Figure I.24. It can be shown that the interference signal 2, which in Figure I.25 overlaps the useful signal 3 (upper sideband), is in fact suppressed. Quadrature mixing of the useful signal $S_{RX}(t) = \sin(\omega_3 \cdot t)$ and the interference signal $N_{RX}(t) = \sin(\omega_2 \cdot t)$ with the LO frequency ω_{LO} produces the following products behind the low-pass filter in the I path or behind the phase shifter in the Q path:

$$S_I(t) = \sin((\omega_3 - \omega_{LO}) \cdot t)$$
$$S_Q(t) = \cos((\omega_3 - \omega_{LO}) \cdot t - 90°) \tag{I.4}$$

where
$S_I(t)$ = real component of the useful signal at time (t), in V
$S_Q(t)$ = imaginary component of the useful signal at time (t), in V
ω_3 = angular frequency of the useful signal, in rad/s
ω_{LO} = angular frequency of the LO injection signal, in rad/s
t = considered time, in s

$$N_I(t) = \sin((\omega_2 - \omega_{LO}) \cdot t)$$
$$N_Q(t) = \cos((\omega_2 - \omega_{LO}) \cdot t + 90°) \tag{I.5}$$

where
$N_I(t)$ = real component of the interference signal at time (t), in V
$N_Q(t)$ = imaginary component of the interference signal at time (t), in V
ω_2 = angular frequency of the interference signal, in rad/s
ω_{LO} = angular frequency of the LO injection signal, in rad/s
t = considered time, in s

Substituting $(\omega_3 - \omega_{LO}) \cdot t = x$ and $(\omega_2 - \omega_{LO}) \cdot t = -y$ ($-y$ is negative when the time is positive, because $f_2 < f_0$), it follows that

$$S_I(t) = \sin(x)$$
$$S_Q(t) = \cos(x - 90°) = \sin(x)$$
$$N_I(t) = \sin(-y) = -\sin(y)$$
$$N_Q(t) = \cos(-y + 90°) = \sin(y)$$

Finally, the summation of I path and Q path results in the AF signals:

$$S(t) = S_I(t) + S_Q(t) = 2 \cdot \sin((\omega_3 - \omega_{LO}) \cdot t) \quad (\rightarrow \text{useful signal with double amplitude})$$

$$N(t) = N_I(t) + N_Q(t) = 0 \quad (\rightarrow \text{interference is suppressed})$$

It can be seen that the condition $N(t) = 0$ is met only if the two components $N_I(t)$ and $N_Q(t)$ have exactly the same amplitude and a phase of 180°. A deviation of only 0.1 dB in the amplitude or 1° in the phase decreases the suppression of the interfering sideband to only

45 dB. With conventional analog signal processing in a direct mixer, sufficiently small tolerances can be obtained only with complex circuitry and arduous tuning, especially when covering a wide frequency band.

I.3.4 Accuracy and Reproducibility

The strong points of digital processing are high accuracy and reproducibility. When using digital circuits for the functional blocks of local oscillator, mixer, filter, and phase shifter only the resolution (number of bits) affects the accuracy. With a 24 bit DSP an attenuation figure of the unwanted sideband of more than 100 dB can be achieved. In fact, floating point processors for real-time processing are available with 32 bits and more [23]. Figure I.26 shows the passband characteristic of such an SSB filter. Without the complex signal processing described above, there would virtually be a second receive channel. The filter was designed in the finite impulse response (FIR) structure with 256 taps according to [24]. It shows a very good shape factor (see Fig. III.42) of better than 1.2 (-6 dB/-60 dB) and has a constant phase of $0°$ for the I channel and $90°$ for the Q channel (Fig. I.24) in its passband. For this reason, the $90°$ phase shifter of the Q channel can be omitted.

The frequency response in Figure I.27 corresponds to a 7 kHz filter designed and optimized for AM reception. With little frequency separation from the limit frequencies the attenuation figure for the cutoff region is already approximately 105 dB.

Another advantage of the digital solution is the fact that such filters can be produced in large quantities with high precision, while there is no aging or drift with temperature variations.

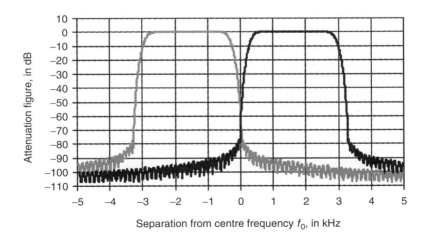

Figure I.26 Measured frequency response of a 2.7 kHz filter for receiving class J3E emission for demodulating the upper or lower sideband. The passband characteristics are fully symmetrical and provide a close-in selectivity (Section III.6) that is clearly above 90 dB already in low separation to the limit frequency.

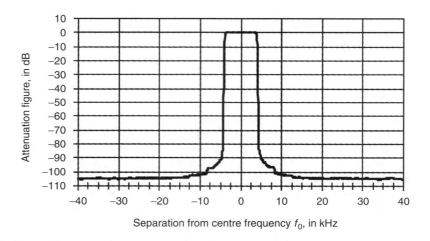

Separation from centre frequency f_0, in kHz

Figure I.27 Measured frequency response of a 7 kHz filter for receiving class A3E emission. No selection gaps were found. The shape factor is below 1.2 and the effective passband ripple (Fig. III.42) is in the range of about 0.3 dB.

I.3.5 VFO for Frequency Tuning

Another important component is the numerically controlled oscillator (NCO). Its construction follows the design of DDS generators (Section I.2.2) used in newer transceivers, but in contrast to these does not require a D/A converter. DDS components with moderate resolution generate a high amount of spurious signals (Section III.7.2). They are not suitable for use as variable frequency oscillators (VFOs), even though the frequency range would be suitable. This is because the low resolution of the D/A converter, ranging from only 8 to 12 bits while signal processing, is carried out with 32 bits. In an all-digital receiver there is no need to change to analog signals. The mixer, which is actually a digital down-converter (DDC), can be controlled with a resolution of 20 bits without problem. A DDS generator with a 10 bit D/A converter has a spurious signal ratio of about 55 dB. Owing to its doubling to 20 bits, a ratio of 110 dB is to be expected. Spurious signals are therefore negligible.

Figure I.28 illustrates the operating principle of the NCO. In the accumulator an increment is added to the value of the 32 bit wide sum register, and the resulting value is stored with every clock cycle in the sum register. The value increases linearly with every clock period until the register overflows at $\geq 2^{32}$. The result is a sawtooth signal having the frequency

$$f_{\text{NCO}} = f_{\text{cl}} \cdot \frac{s}{2^N} \tag{I.6}$$

where
f_{NCO} = NCO output frequency, in Hz
$\quad f_{\text{cl}}$ = clock frequency, in Hz
$\qquad s$ = decimal value of the $(N{-}1)$ bit wide control increment, dimensionless
$\quad N$ = word length of the phase accumulator, in bits

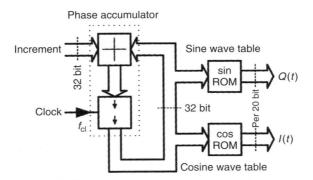

Figure I.28 Principle of the digital oscillator (NCO) for shifting the received signal by 0 Hz. The adder functions together with the sum register as a phase accumulator to which a fixed increment (the control word s) is added with a clock frequency of f_{cl}.

and an achievable frequency resolution [25] of

$$\Delta f_{NCO} = \frac{f_{cl}}{2^N} \tag{I.7}$$

where
Δf_{NCO} = achievable frequency resolution of the NCO, in Hz
$\quad f_{cl}$ = clock frequency, in Hz
$\quad N$ = word length of the phase accumulator, in bits

A table stored in a non-volatile memory (ROM – read only memory) is used for converting the sawtooth wave to a sine or cosine signal pattern in 2^{20} steps, which determine the possible phase resolution of the output signal:

$$\Delta \phi_{NCO} = \frac{2 \cdot \pi}{2^M} \tag{I.8}$$

where
$\Delta \phi_{NCO}$ = achievable phase resolution of the NCO, in rad
$\quad M$ = number of address bits of the ROMs, in bits

This essentially influences the number and spectral separation of the spurious signals from the useful signal [25].

With a clock frequency of 65 MHz and, for example, a decimal value of 231,267,470 as increment in the sum register the NCO generates an output frequency of

$$f_{NCO} = 65 \text{ MHz} \cdot \frac{231,267,470}{2^{32 \text{ bit}}} = 3.5 \text{ MHz}$$

A deviation of 1 to 231,267,471 causes a frequency alteration of 0.015 Hz. This shows that it is possible to tune the frequency with a resolution as high as

$$\Delta f_{NCO} = \frac{65 \text{ MHz}}{2^{32 \text{ bit}}} = 15 \text{ mHz}$$

and a phase resolution of

$$\Delta\phi_{NCO} = \frac{2\cdot\pi}{2^{20\,bit}} = 5.99 \ \mu rad \ \hat{=} \ \frac{360°}{2^{20\,bit}} = 0.000,343°$$

Compared with the conventional analog design the digital design has the following additional advantages:

- The frequency achieved is as stable as the quartz crystal.
- Large frequency jumps can be made within microseconds and with extremely short transient periods (Section III.15) (for e.g. in spread-band technology applications).
- With well-considered dimensioning and the corresponding width of the ROMs the sideband noise, and therefore reciprocal mixing (Section III.7) is very low.
- There is not any response of the NCO to the receiver input, so that there is no stray radiation from the receiver (Section III.17).

I.3.6   Other Required Hardware

A possible circuit design for realizing the functional blocks described will be described based on the example of the unit shown in Figures I.21, I.22 and I.29. Analog Devices' IC AD6624 down-converter has been chosen for performing the functions of mixer, NCO, decimation and filtering [26]. In addition a digital signal processor is required. The same manufacturer's model ADSP-21362 already described is suitable for this purpose [23]. This DSP will perform the following functions:

- Filtering and near selection of the receive signal (several receive bandwidths from 50 Hz to 25 kHz, see Figs. I.26 and I.27).
- Measuring the receive signal voltage for the S meter (Section III.18.1) and the automatic gain control (AGC).
- IQ demodulation.
- Communication with the audio CODEC for the connection of loudspeakers, headphones or media for analog AF recording.
- Controlling the digital down-converter (configuration, frequency tuning, AGC).
- Communication with the operating unit or control device (PC or keyboard, display, or incremental encoder for frequency tuning).
- In addition, other functions, like adaptive noise suppression, modem functions, demodulation of coded modulation modes, and the functions of CW decoder or terminal node controller (TNC).

Figure I.30 shows a block diagram of the resulting all-digital VLF/HF communications receiver. The chipset consists of only four blocks. Almost unbelievable is the drastic reduction in the amount of hardware compared with a unit of conventional circuit design having the same characteristics.

The spurious-free dynamic range (Section I.3.2) of the AD6645 A/D converter used can be increased up to 100 dB by dithering (Section III.9.5). In the course of development the use of a noise source has been investigated. However, tests have shown that in

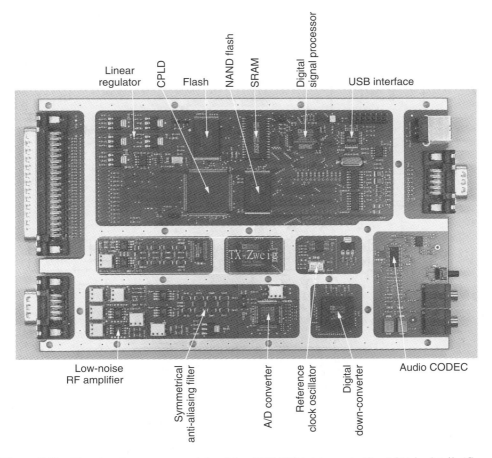

Figure I.29 The signal processor module of the ADT-200A (shown in Fig. I.21) in detail. (Company photograph of Hans Zahnd engineering consultants.)

practical applications there are always a sufficiently high number of stochastic interference signals, so that the additionally generated noise brings no further improvement except in intermodulation measurements (Section III.9.10) using the established two-tone measuring method (Fig. I.31).

I.3.7 Receive Frequency Expansion by Subsampling

The Nyquist sampling theorem specifies that a signal can be correctly reconstructed only if the sampling frequency (f_s) is at least twice as high as the highest frequency component of the sampled signal. If this condition is not met, aliasing of the frequencies above $f_s/2$ takes place in the region below $f_s/2$. For example, with $f_s/2 = 32.5$ MHz a receive frequency of 35 MHz will be superimposed on a receive frequency of 30 MHz. When increasing the frequency the process is repeated in segments of $f_s/2$. These segments are

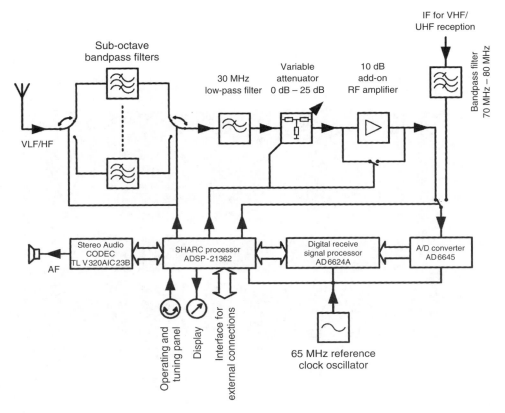

Figure I.30 Block diagram of the (fully) digitized radio receiver in the ADT-200A (Fig. I.21). The frontend, consisting of the bandpass filters and the 30 MHz low-pass filter, is constructed in keeping with present day technology. Signal processing is done by a chipset comprising four highly integrated components of the latest generation. In the VLF/HF receiving range the unit functions as a digital RX of generation 2.5 and for receiving frequencies in the VHF/UHF range as a second-generation digital RX by including subsampling (Section I.3.7).

called Nyquist windows (Fig. I.32). The phenomenon can be useful for receive frequency ranges above the sampling frequency, provided that they do not exceed the segment limits. A suitable A/D converter can use a low sampling frequency and still cover a range of several hundred MHz. This is called *subsampling*. The upper frequency limit is determined by the uncertainty in the sampling circuit of the A/D converter, which is called aperture jitter. Also, the phase noise (Fig. III.49) of the sampling frequency becomes more relevant with an increase in the signal frequency. (The implementation of sigma delta A/D converters for sampling such bandpass-filtered frequency ranges is advantageous [9]. How the bandpass subsampling [27] can be performed with a sigma delta A/D converter is investigated in [28].)

Generation of the intermediate frequency as shown in the block diagram in Figure I.30 for VHF/UHF reception is according to this principle. The IF is in the rather unusual range

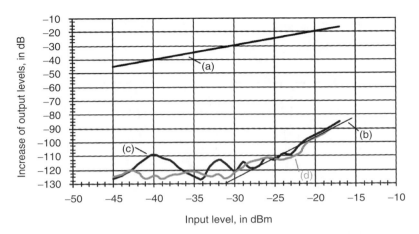

Figure I.31   Actual intermodulation response of the third order (Section III.9.2) as used in the all-digital radio receiver described. Curve a) represents the power of one of the excitation signals fed in. The level increase of an IM3 product without dithering is indicated by curve c), and with active dithering by curve d) (Section III.9.5). Especially with the received levels under normal operating conditions the dithering function brings a significant improvement in the intermodulation immunity (Section III.9.6). The IM3 increase expected by definition is shown in curve b) for comparison. (The increase in the input level of over -25 dBm is caused by intermodulation in the analog frontend of the unit. This is the reason for the level increase of 3 dB per 1 dB increase in the excitation signal.)

between 70 MHz and 80 MHz, which allows the use of a simple transverter with a fixed heterodyne frequency, thus providing high image frequency rejection (Section III.5.3) even with moderate input selection properties. If an IF signal is sampled by means of bandpass subsampling, the desired spectrum is mirrored at $f_s/4$ [9]. With subsampling the range 70 MHz to 80 MHz is shifted downward to 5 MHz to 15 MHz (Fig. I.32).

I.4   Practical Example of a Portable Wideband Radio Receiver

Possible implementations of modern wideband receivers covering a wide(r) receive frequency range differ from previous designs in several respects, both in regard to the basic parameters covered (frequency range, demodulated class(es) of emission, intended use, etc.) and in the specific circuit layout. Most of these are designed as multiple-conversion superhets (Section I.2.2) with a high first intermediate frequency. Depending on the required technical properties and, particularly their flexibility in terms of equipment configuration, they often operate with digital signal processing from the IF stage further downstream with a lower frequency.

To discuss this present state-of-the-art design, the unit shown in Figure I.33 can be taken as an example [29], and [30]. This unit covers the receive frequency range continuously from 9 kHz to 7.5 GHz. Despite its small dimensions it also offers a wide range of functions while being highly mobile at the same time. The moderate power consumption allows prolonged line-independent operation from a rechargeable lithium ion battery pack.

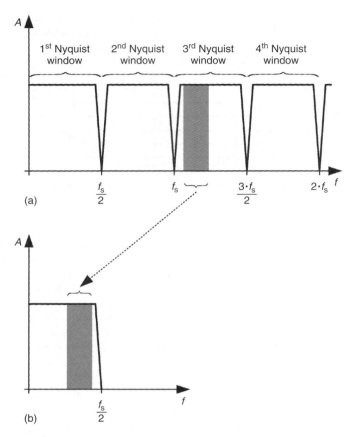

Figure I.32 Expanding the input frequency range by subsampling. Graph (a) shows the Nyquist windows and graph (b) the shifted frequency segment that was initially above the receive frequency range. The subsampling frequency range fed to the A/D converter must have a bandwidth of $<f_s/2$.

Its graphic display of frequency occupancy and analysis of receive signals extends the functionality compared with units designed for demodulation alone.

I.4.1 Analog RF Frontend for a Wide Receive Frequency Range

The signals received at the antenna port pass a low-pass filter, limiting the frequency spectrum for further processing to the filter's limit frequency of 8 GHz. Subsequent signal processing is carried out in three different paths, depending on the receive frequency selected (Figs. I.34 and I.35):

- Signals in the frequency range between 9 kHz and 30 MHz are fed via a 30 MHz low-pass filter and a HF preamplifier directly to the A/D converter. In a multi-functional portable unit of the given dimensions (Fig. II.41) a selective frontend selection of this

Figure I.33 PR100 allows continuous tuning for receiving in a frequency range from 9 kHz to 7.5 GHz with a noise figure (Section III.4.2) of less than 20 dB across the entire receive frequency range. The portable unit has a weight of 3.5 kg, including the battery pack (see Fig. II.41). Analysis of the received signals can also be carried out on the 6.5 inch colour display. Information can be stored on a built-in SD memory card without external accessories. For special applications all receiver functions can be remotely controlled via a LAN interface. The unit can be upgraded optionally for use as a single-channel direction finder in the range from 20 MHz to 6 GHz (Company photograph of Rohde&Schwarz).

frequency spectrum with its high levels is not possible. With low receive frequencies below 30 MHz the unit therefore operates as a direct receiver (Section I.3).
- In the frequency range between 20 MHz and 3.5 GHz the signal passes several automatically activated bandpass filters of moderate quality under operating conditions (Section III.11) or a high-pass filter and a subsequent RF preamplifier. For high-level input signals an attenuator allows bypassing of the preselector and RF preamplifier to prevent the generation of high sum signals and to ensure that the first IF stage operates in the linear region of its dynamic range. It forms the front block of the IF-generating circuit to which the filtered or attenuated input spectrum is fed via a 3.5 GHz low-pass filter.
- Signals in the frequency range above 3.5 GHz to 8 GHz are fed to an IF generating circuit via a high-pass filter of 3.5 GHz limit frequency and an RF preamplifier followed by another 8 GHz low-pass filter.

The above-mentioned IF-generating circuitry converts the respective receive frequency band to three intermediate frequencies, of which the last analog third IF is 21.4 MHz (the analog unregulated frequency of 21.4 MHz is available at a BNC socket for external processing). Following this signal preparation the receive frequency range from 20 MHz to nearly 8 GHz can now be fed to the same A/D converter as the lower-frequency receive signals. At higher frequencies, the concept of the multiple-heterodyne receiver with A/D conversion after the third IF stage is therefore used (see Fig. I.16). To make these specific equipment parameters available [30], the subsequent stages effectively process signals only up to 7.5 GHz.

I.4.2 Subsequent Digital Signal Processing

From the A/D converter on, the signals conditioned as shown in Figure I.34 can be processed in the downstream functional blocks according to the principles outlined in Sections I.3.1 to I.3.6.

In order to use the portable wideband radio receiver shown in Figure I.33 for graphic evaluations and analyses in addition to demodulation, the signal path is divided after the A/D converter into two parallel branches (Figs. I.36 and I.37). This allows simultaneous demodulation, receive signal level measurement (Section III.18), and display of a spectral panorama [29]. These two branches are described in detail in the next two sections.

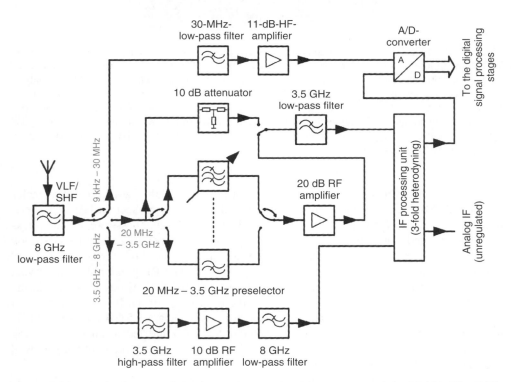

Figure I.34 Block diagram of the frontend up to the A/D converter of the PR100 (Fig. I.33). The receive frequency range is extended by a combination of second-generation digital RX (for receiving frequencies above 30 MHz) and of generation 2.5 (for frequencies below 30 MHz). For input signals >20 MHz the analog unregulated signal is available externally via the 41.4 MHz IF (subsequent digital processing paths are illustrated in Figures I.36 and I.37).

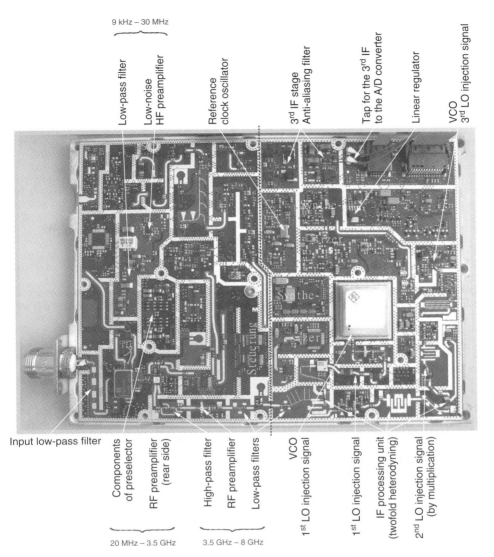

Figure I.35 Front detail of the analog RF frontend module of the PR100 (shown in Fig. I.33; see also Fig. I.34). The entire half to the left of the dotted line contains the preselection of the three receive paths, while the right half includes the frequency processing and the IF paths. (Company photograph of Rohde&Schwarz.)

I.4.3 Demodulation with Received Signal Level Measurement

The signal is prepared for demodulation or level measurement (Fig. I.36) by a digital down-converter (DDC) (Section I.3.5) and a digital bandpass filter. For the matched reception of different classes of emission and for an optimized signal-to-interference ratio in different receiving situations, the receiver offers the possibility of choosing from 15 digitally realized IF filter bandwidths from 150 Hz up to 500 kHz (in part depending on

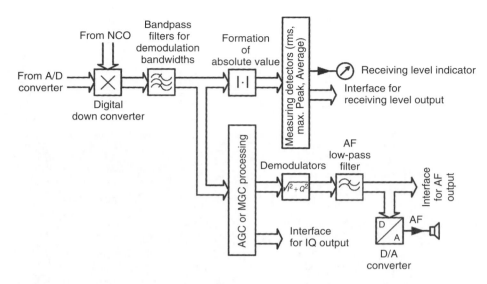

Figure I.36 Execution of the digitally based demodulation stage (lower signal path) and measuring the level of the receive signal (upper signal path) (the frontend up to the A/D converter is shown in Figure I.34).

the emission class selected). These can be selected independently of the display range and the resolution bandwidth of the spectral display described in Section I.4.4.

For demodulating analog signals the complex baseband data (Section I.2.3) are fed via the bandpass filter to the AGC or MGC stage (Fig. I.5). They are then subjected to the selected demodulation algorithm for A1A (CW), A3E (AM), B8E (ISB), F3E (FM), J3E (SSB, upper and lower sideband), or pulse. The results are in digital form and made available via the LAN interface. For loudspeaker output the digital audio data stream must be converted back to an analog signal.

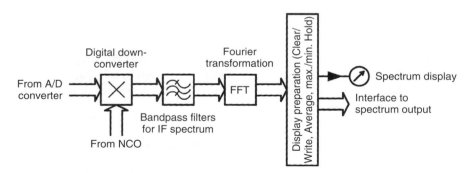

Figure I.37 Structure of the digital path for displaying the receive signal spectrum from the intermediate frequency by FFT analysis. Displaying the signal levels requires comprehensive logarithmic calculations of all the bins (the frontend up to the A/D converter is shown in Figure I.34).

After the AGC or MGC stage the complex IQ data (Section I.2.3) of the digital signals are directly available for further processing.

For the purpose of measuring the receive signal level, the signal strength is determined and the value assessed according to the measuring detector selected (rms, maximum peak, sample, average, as known from spectrum analyzers). The measured and evaluated levels are then available on the display and at the LAN interface. In addition, the unit is able to refer to a set of internally stored correction factors that enable measurement of the field strengths for known antenna factors (Section III.18).

I.4.4   Spectral Resolution of the Frequency Occupancy

The second signal path with DDC and digital bandpass filter is used for the calculation of the signal spectrum around the receive frequency in the FFT block (Fig. I.37) from the intermediate frequency. The bandwidth of the bandpass filter and, with it, the associated spectral span on the display can be selected by the user in the range from 1 kHz up to a maximum of 10 MHz.

The calculations based on the fast Fourier transformation (FFT) of the IF-filtered data stream (Section II.4.2) have a considerable advantage: the receiver sensitivity and signal resolution are clearly superior to those of a conventional analog receiver with the same spectral display span.

When selecting the setting of, for example, 10 kHz for sensitive signal reception, the following steps are performed in the course of the FFT calculation. Based on the finite steepness (Fig. III.42) of the IF filter, the decimated sampling rate (Section I.3.2) must be higher than the selected display width. This means that the quotient of the decimated sampling rate and bandwidth is >1 and represents a measure of the steepness of the IF filter (this may be seen as similar to the shape factor described in Section III.6.1). Its numerical value depends on the display range selected and may vary. For the 10 kHz display span, the constant is 1.28 and results in the necessarily decimated sampling rate of 10 kHz · 1.28 = 12.8 kHz. The FFT standard length n of the unit described in Figure I.33 is 2,048. The calculation (with a Blackman window) divides the frequency band of 12.8 kHz into 2,048 equidistant FFT lines (also called frequency lines or bin widths). Each of these FFT lines represents a quasi receive channel with a resolution bandwidth of

$$B_{res}(n) = \frac{f_s}{n} \qquad (I.9)$$

where
B_{res} = resolution bandwidth of an FFT line, in Hz
f_s = decimated sampling rate prior to FFT analysis, in Hz
n = number of FFT lines, dimensionless

For the display range considered, this results in a resolution bandwidth of

$$B_{res}(n = 2{,}048) = \frac{12{,}800 \text{ Hz}}{2{,}048} = 6.25 \text{ Hz}$$

per frequency line, roughly effective as an equivalent noise bandwidth (Section III.4.4), and corresponds to

$$B_{\text{dB N}} = 10 \cdot \lg \left(\frac{6.25 \text{ Hz}}{1 \text{ Hz}} \right) = 8 \text{ dBHz}$$

This allows the determination of the minimum discernible signal (Section III.4.5) (the noise floor) of the spectral display using Equation (III.10):

$$P_{\text{MDS}}(B_{-6 \text{ dB}} \approx 6.25 \text{ Hz}) = -174 \text{ dBm/Hz} + 20 \text{ dB} + 8 \text{ dBHz} = -146 \text{ dBm}$$

According to specification [30], with some receive frequencies the noise figure (Section III.4.2) of the receiver is below the value of 20 dB used for the calculation (in some cases below 10 dB), which suggests an even better sensitivity within this frequency ranges.

The minimum display range above 1 kHz results in the maximum sensitivity, while the widest range of 10 MHz produces the lowest sensitivity. The high spectral resolution of the FFT calculation shows that closely adjacent signals appear well separated in the spectrum displayed.

Prior to feeding the IF spectrum to the display or LAN interface, the type of display is prepared according to the user's specification (normal or clear/write, average, max. hold, min. hold) (as is also known from spectral analyzers).

For a survey over a wider spectral panorama several of the up to 10 MHz wide FFT display ranges can be combined on the frequency axis to form wide display ranges (so-called panorama scans, Figs. II.23 and II.24). In this operating mode the user can choose from 12 bin widths between 120 Hz and 100 kHz. Based on the selected bin width and the start and stop frequency settings, the required FFT length and the width of the frequency window for each individual viewing increment are determined automatically. However, the panorama scan must be stopped when operating the receiver in the listening mode [30].

References

[1] Olaf Koch: Hochlineare Eingangsmischer für Kurzwellenempfänger (Highly Linear Input Mixers for Short-Wave Receivers); manuscripts of speeches from the Short-Wave Convention Munich 2001, pp. 91–105

[2] Hans H. Meinke, Friedrich-Wilhelm Gundlach editors: Taschenbuch der Hochfrequenztechnik (Handbook for Radio Frequency Technology), 5th edition; Springer Verlag 1992; ISBN 3-540-54717-7

[3] Erich H. Franke: Fractional-n PLL-Frequenzsynthese (Fractional-n PLL Frequency Synthesis); manuscript of speeches from the VHF Convention, Weinheim 2005, pp. 7.1–7.10

[4] Analog Devices, publisher: Datasheet 1 GSPS Direct Digital Synthesizer with 14-Bit-DAC AD9912; Rev. 0/2007

[5] Markus Hufschmid: Empfängertechnik (Receiver Technology); manuscript from the FH Nordwestschweiz 2008, pp. 1–14

[6] Thomas Valten: Digitale Signalverarbeitung in der Kurzwellen-Empfängertechnik (Digital Signal Processing in Short-Wave Receiver Designs); manuscripts of speeches from the Short-Wave Convention Munich 2001, pp. 69–80

[7] Rüdiger Leschhorn, Boyd Buchin: Software-basierte Funkgeräte – Teil 1 und Teil 2 (Software-Based Radio Equipment – Part 1 and Part 2); Neues von Rohde&Schwarz II/2004, pp. 58–61, Neues von Rohde&Schwarz III/2004, pp. 52–55; ISSN 0548–3093

[8] Hans Zahnd: Software Radio – Technologie der Zukunft (Software Radio – Technology of the Future); CQ DL 8/2000, pp. 580–584; ISSN 0178-269X

[9] Anne Wiesler: Parametergesteuertes Software Radio für Mobilfunksysteme (Parameter-Controlled Software Radio for Mobile Radio Systems); research reports from the Communications Engineering Lab of the Karlsruhe Institute of Technology, Vol. 4/2001; ISSN 1433–3821

[10] Arnd-Ragnar Rhiemeier: Modulares Software Defined Radio (Modular Software-Defined Radio); research reports from the Communications Engineering Lab of the Karlsruhe Institute of Technology, Vol. 9/2004; ISSN 1433–3821

[11] Eric T. Red: Digitale Empfänger – weiterhin Zukunftsmusik? (Digital Receivers – A Futuristic Vision any Longer?); beam 9/1988, pp. 26–31; ISSN 0722–0421

[12] Thomas Rühle: Entwurfsmethodik für Funkempfänger – Architekturauswahl und Blockspezifikation unter schwerpunktmäßiger Betrachtung des Direct-Conversion- und des Superheterodynprinzipes (Methodology of Designing Radio Receivers – Architecture Selection and Block Specification with the Main Focus on Direct Conversion and Superheterodyne Designs); dissertation at the TU Dresden 2001

[13] Burr-Brown, publisher: Preliminary Information 14-Bit – 65 MHz Sampling ANALOG-TO-DIGITAL Converter ADS852; Rev. 6/1998

[14] Analog Devices, publisher: Datasheet 14-Bit – 80 MS/s/105 MS/s A/D Converter AD6645; Rev. C/2006

[15] Analog Devices, publisher: Datasheet 16-Bit – 80/100 MS/s ADC AD9446; Rev. 0/2005

[16] Linear Technology, publisher: Datasheet 14-Bit – 130 MS/s ADC LTC2208-14; Rev. A/2006

[17] Eugene B. Hogenauer: An economical class of digital filters for decimation and interpolation; IEEE Transactions on Acoustics, Speech and Signal Processing 2/1981 – Vol. 29, pp. 155–162; ISSN 0096–3518

[18] Peter E. Chadwick: HF Receiver Dynamic Range – How Much Do We Need?; QEX 5+6/2002, pp. 36–41; ISSN 0886–8093

[19] Herbert Steghafner: Breitbandempfänger (Broadband Receiver); Rohde&Schwarz 2000; patent application DE10025837A1

[20] Pervez M. Aziz, Henrik V. Sorensen, Jan van der Spiegel: An Overview of Sigma-Delta Converters; IEEE Signal Processing Magazine, 1/1996 – Vol. 13, pp. 61–84; ISSN 1053–5888

[21] Feng Chen, B. Leung: A 0.25-mW Low-Pass Passive Sigma-Delta Modulator with Built-In Mixer for a 10-MHz IF Input; IEEE Journal of Solid-State Circuits 6/1997 – Vol. 32, pp. 774–782; ISSN 0018–9200

[22] Shengping Yang, Michael Faulkner, Roman Malyniak: A tunable bandpass sigma-delta A/D conversion for mobile communication receiver; Proceedings of IEEE 44th Vehicular Technology Conference Stockholm 1994 – Vol. 2, pp. 1346–1350; ISSN 1090–3038

[23] Analog Devices, publisher: Datasheet SHARC Processors ADSP-21362/ADSP-21363/ADSP-21364/ADSP-21365/ADSP-21366; Rev. C/2007

[24] Rob Frohne: A High-Performance, Single-Signal, Direct-Conversion Receiver with DSP Filtering; QST 4/1998, pp. 40–45; ISSN 0033–4812

[25] Anselm Fabig: Konzept eines digitalen Empfängers für die Funknavigation mit optimierten Algorithmen zur Signaldemodulation (Concept of a Digital Receiver for Radio Navigation Using Optimized Algorithms for Signal Demodulation); dissertation at the TU Berlin 1995

[26] Analog Devices, publisher: Datasheet Four-Channel 100 MS/s Digital Receive Signal Processor (RSP) AD6624A; Rev. 0/2002

[27] Friedrich K. Jondral: Die Bandpassunterabtastung (Bandpass Subsampling); AEÜ 1989 – Vol. 43, pp. 241–242; ISSN 1434–8411

[28] Brenton Steele, Peter O'shea: A reduced sample rate bandpass sigma delta modulator; Proceedings of the Fifth International Symposium on Signal Processing and its Applications 1999 – Vol. 2, pp. 721–724, ISBN 1-86435-451-8

[29] Peter Kronseder: Funkstörungen optimal erfassen – Portabler Monitoring-Empfänger für Signalanalysen von 9 kHz bis 7.5 GHz (Optimum Detection of Radio Interferences – Portable Monitoring Receiver for Analyzing Signals from 9 kHz to 7.5 GHz); www.elektroniknet.de 7/2008

[30] Rohde&Schwarz, publisher: Datenblatt Tragbarer Empfänger R&S®PR100 – Portable Funkerfassung von 9 kHz bis 7.5 GHz (Datasheet on Portable Receiver R&S®PR100 – Portable Radio Signal Detection from 9 kHz to 7.5 GHz); Rev. 01.02/2008

Further Reading

Analog Devices, publisher: Datasheet 12-Bit – 65 MS/s IF Sampling A/D Converter AD6640; Rev. A/2003

Analog Devices, publisher: Datasheet 14-Bit – 40 MS/s/65 MS/s Analog-to-Digital Converter AD6644; Rev. D/2007

Friedrich K. Jondral: Kurzwellenempfänger mit digitaler Signalverarbeitung (Short-Wave Receiver Using Digital Signal Processing); Bulletin of the Schweizerische Elektrotechnischer Verein 5/1990 – Vol. 81, pp. 11–21; ISSN 036–1321

Joe Mitola: The Software Radio Architecture; IEEE Communications Magazine 5/1995 – Vol. 33, pp. 26–38; ISSN 0163–6804

Ulrich L. Rohde, Jerry Whitaker: Communications Receivers; 2nd edition; McGraw-Hill Companies 1997; ISBN 0-07-053608-2

James Scarlett: A High-Performance Digital-Transceiver Design – Part 1; QEX 7+8/2002, pp. 35–44; ISSN 0886–8093

Texas Instruments, publisher: Data manual Stereo Audio CODEC, 8 to 96 kHz, with Integrated Headphone Amplifier TLV320AIC23B; Rev. H/2004

II

Fields of Use and Applications of Radio Receivers

II.1 Prologue

Receivers are used in a wide range of applications independently of their type of construction (Fig. I.20). The descriptions in the following paragraphs will focus on terrestrial applications. In general, the main goals defined in performance specifications are:

- 'most cost-effective designs for the mass market (consumer electronics)'

through,

- 'higher technical demands regarding specific design parameters or receiver characteristics'

up to,

- 'highly sophisticated special or general-purpose units to meet the highest commercial or military demands regarding both, receiver characteristics (Part III) and the sturdiness of electronics, mechanics, and other equipment parts.' (Such equipment will be discussed in detail in the text below, as this information is scarce in the common literature.)

The *collective name radio receiver* refers to a design for the reception of wireless transmissions based on the utilization of electromagnetic waves (Fig. II.1). Ideally, this device extracts the full information content from the incident signal. Regarding wireless transmission technology, the *fields of use* for such devices can be divided into two main groups: units for receiving information/messages and units for measuring purposes. Over time, many different terms have been used which, today, become more and more blurred and may be summarized under the following main groups.

Communications receivers are intended for information retrieval from general or specific emissions (see Table II.5) received via an antenna. These units enable the reception of certain transmitting channels or frequency ranges used for the respective class of emission or

Radio Receiver Technology: Principles, Architectures and Applications, First Edition. Ralf Rudersdorfer.
© 2014 Ralf Rudersdorfer. Published 2014 by John Wiley & Sons, Ltd.

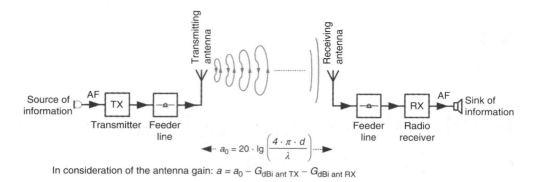

In consideration of the antenna gain: $a = a_0 - G_{\text{dBi ant TX}} - G_{\text{dBi ant RX}}$

Figure II.1 Wireless transmission path with its basic elements. The electromagnetic wave front emitted from the transmitting antenna impinges on the receiving antenna after unhindered propagation (free first Fresnel zone) reduced by the path loss (also called free-space attenuation or spreading loss). For calculations of the signal intensity received with antennas that provide a signal gain, the path loss must be reduced by the antenna gain figure in direction of the wave propagation. (a_0 = free-space attenuation figure, in dB; $\pi = 3.1416$; d = antenna distance, in m; λ = wavelength, in m; $G_{\text{dBi ant TX}}$ = transmitting antenna gain figure, in dBi; $G_{\text{dBi ant RX}}$ = receiving antenna gain figure, in dBi)

modulation types. External criteria of particular importance are simple operation, optimally adapted to the intended use, and a sufficiently high quality of 'information retrieval' (Fig. II.2).

Measuring/test receivers are intended for the (often standardized) measurement and evaluation of electromagnetic radiation/transmission, of (radio) interferences, or of the parameters of the signals to be transmitted. The accuracy of the measurement is a very important characteristic of such receivers.

Another commonly used collective name is *short-wave receiver*. This term has historical roots, and in the currently still used segmentation of the frequency spectrum into a range below 30 MHz and another range of 30 MHz and above. Owing to the many natural characteristics of wave propagation it is necessary that radio receivers using frequencies up to 30 MHz meet specific requirements. (By definition a short-wave receiver covers the short-wave frequency range from 3 MHz to 30 MHz.) For practical purposes the term short-wave receiver is a designation used colloquially for almost all equipment operating with a more or less extended receive frequency range below 30 MHz. This applies especially to units designed as all-wave receivers (Section II.3.2) for frequencies of up to 30 MHz (Fig. II.3). The term short-wave receiver is often used for those units described in Sections II.3.2, II.3.3 and II.3.4 as well as in Sections II.4 and II.6, independently of or in addition to more descriptive names.

II.2   Wireless Telecontrol

In the technical field of wireless telecontrol large numbers of simple receivers are used. Applications range from decentralized radio remote control of industrial machinery and

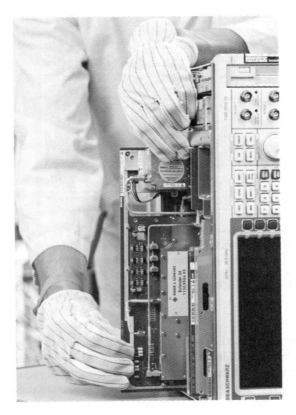

Figure II.2 Example of a service-friendly mechanical construction for a radio receiver with the individual circuit boards plugged into the chassis. (Company photograph of Rohde&Schwarz.)

Figure II.3 Best possible screening and highest crosstalk attenuation is achieved by a professional modular concept: Individual modules are seen as plug-in units in the 19″ chassis. Critical signals are conducted between the individual modules by screened coaxial cables in 50 Ω technology.

Figure II.4 Simple low-cost two-channel receiver for telecontrol purposes in the 35 cm ISM band. Two highly durable relays in the output circuit allow the potential-free shifting of the loads. Typical applications are wireless operation of garage doors, awnings, lights or fountains in ponds. For each output an automatic switch-off time can be programmed and random wire antennas are used. The systems are operated with a hand-held transmitter, the remote control. (Company photograph of RosyTec.)

automatic door openers (Fig. II.4) to reading measurements of remotely located sensors. Furthermore, they are found in remote switches (e.g., keyless car entry) or in audio data transfer to wireless headphones and loudspeakers in the home environment. For this purpose, standard technologies like Bluetooth [1] and ZigBee are used to some extent, but simple data exchange methods based on types of frequency-keyed and amplitude-keyed modulation using individual transfer protocols or simple frequency modulation are also utilized. In essence, this is directional (either bidirectional or unidirectional, depending on the specific use) information transfer. The operating frequencies are usually in a range that requires no user permit, the so-called ISM frequency bands (industrial scientific medical bands) (Table II.1). Transmission systems of this type are also called short-range devices (SRD).

Today, the actual *data receiver* is usually a fully integrated component tailored to its specific use. This contains the entire *single-chip receiver* (Figs. II.5 and II.50) and is very cost-effective, but often has limited technical capabilities (Part III). More details about the actual design and the procedures of the dimensioning of such receiver components can be found in [3].

II.2.1 Radio Ripple Control

Ripple control is used by power companies to transmit control commands to a large number of customer-premises equipment (CPE). In this way tariff meters, street lighting, loads, etc. can be remotely controlled. Wireless ripple control systems are comprised of the user operating station for issuing individual customer commands, a mainframe

Table II.1 ISM frequency bands according to ITU RR [2]

Band	Range	Frequency
HF	44 m	6,765 kHz–6,795 kHz*
HF	22 m	13.553 MHz–13.567 MHz
HF	11 m	26.957 MHz–27.283 MHz
VHF	7 m	40.66 MHz–40.7 MHz
UHF	70 cm	433.05 MHz–434.79 MHz*
(UHF	35 cm	868 MHz–870 MHz*)
UHF	33 cm	902 MHz–928 MHz**
UHF	12 cm	2.4 GHz–2.5 GHz
SHF	5 cm	5.725 GHz–5.875 GHz
SHF	1.2 cm	24 GHz–24.25 GHz
EHF	5 mm	61 GHz–61.5 GHz*
EHF	2.5 mm	122 GHz–123 GHz*
EHF	1.2 mm	244 GHz–246 GHz*

*Deviations/limitations according to country may apply.
**In ITU RR region 2 only.

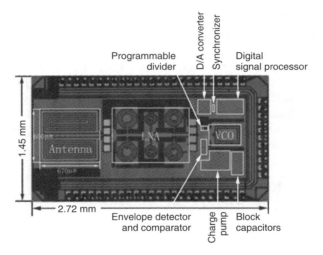

Figure II.5 Fully integrated 5 GHz single-chip receiver, including a wafer-integrated chip antenna and digital baseband-processing with synchronization of the data received. The 0.13 μm CMOS technology enables data rates up to 1.2 Mb/s. (Photograph by the University of Michigan.)

computer, central transmitters for long-wave operation, and radio ripple control receivers. The mainframe computer manages the commands and forwards them to the transmitter as control telegrams at the correct time. These central components of the system (mainframe computer and transmit system) are used by both the power company and the end customer. The computer ensures that the individual transmit demands of each participant are met [4].

The European radio ripple control (EFR – Europäische Funk-Rundsteuerung) repeat all transmissions once automatically and several times optionally. Between the transmissions of the participants, ERF synchronizes the receivers in regard to day and time every 15 s. Thus, the radio ripple control also serves as a time signal transmitter (Section II.7). The two German LW transmitters use an internal carrier frequency of 129.1 kHz (Mainflingen with 100 kW) and 139.0 kHz (Burg with 50 kW). The transmitter in Hungary uses 135.6 kHz (Lakihegy with 100 kW). Throughout Europe, these frequencies are allocated exclusively as ripple control. The three LW transmitters provide good coverage of the German-speaking regions, the Czech Republic, Slovakia, Hungary and the states evolving from the former Yugoslavia. The receive field strength (Section III.18) in the covered area is >55 dB(μV/m). The signal is modulated through frequency shift keying (FSK) with $+170$ Hz. The transmit speed is 200 Bd and corresponds to the transmission format specified in DIN 19244. EFR assigns the addresses to address telegrams and receivers so that each participant can initiate individual telegrams.

On request, a *reference receiver* will report the transmitted telegrams, thus allowing a closed monitoring circuit. In addition to the user operating station the user may also perform remote parameterizing of the radio ripple control receivers via an internet portal or a telecommunications network (e.g., enabling him to reprogram his home heating system or an automatic light switch serving as a burglar deterrence from any given place with the help of his notebook [5]).

The switching telegram received by the *radio ripple control receiver* is read by the integrated decoder and converted to a signal for operating the electromagnetic relays (e.g., for switching a power meter to another energy tariff). Radio ripple control receivers usually use a ferrite antenna and are designed as *fixed-frequency receivers* to operate on a preselected LW transmitter. It is advantageous to use a built-in memory in the radio ripple control receiver to store the switching messages [6]. A basic routing program stored in this program memory will guarantee the automatic execution of all instructions, even in case of a long uninterrupted receiving phase.

II.3   Non-Public Radio Services

Radio receivers or receiving modules of transceivers (radio equipment) are used by all recognized radio services listed in the current edition of [2] and [7] as well as by various other authorized groups that are permitted to use certain frequency channels only (as assigned by the authorities). The receivers are usually designed for demodulating one predetermined class of emission only.

II.3.1   Air Traffic Radio

Aviation radio or aeronautical radio (Figs. II.6 and II.7) uses the frequency range from 117.975 MHz to 137 MHz in conventional amplitude modulation (A3E) with 25 kHz channel spacing. (In addition, the frequency range from 225 MHz to 400 MHz is reserved exclusively for military aircraft radio communication.) In the 'upper airspace' (from FL 245/approximately 24,000 ft/7,500 m) a channel spacing of 8.33 kHz is also used. For wide-area communication in long-distance flights over the sea, where no connection to

Figure II.6 Group of receivers used in a ground control station for communication on the VHF air radio band. A remote-controlled monitoring system (RCMS) enables the technical service to control and monitor individual receivers (by software via a virtual user interface). For increased operational safety two receivers working simultaneously on two antennas per channel are used for communication between aircraft and air traffic control on the ground. Based on the evaluation of the control voltage (Section III.14) the audio frequency of the RX with the highest received voltage (best signal selection – BSS) is forwarded to the air traffic controllers' workplace (refer to Fig. II.43). (Company photograph of Austro Control.)

Figure II.7 Photograph showing details of one of the receivers in Fig. II.6. A power divider splits the signal picked up at the respective receiving antenna and feeds it to a buffer amplifier with a high large-signal immunity. A multiple-circuit preselector (Section III.11) located immediately behind the RX antenna socket relieves the subsequent stages. The operational quality by the depicted model EU231 from Rohde&Schwarz is so high that the unit must be tuned exactly to the receive channel. (Company photograph of Austro Control.)

a ground control station in the VHF range is possible, the short-wave frequencies listed in Table II.2 are also employed in single sideband modulation (J3E) using the upper sideband.

Apart from air traffic radio using conventional voice communication there is also the aeronautical radio navigation service employing the so-called *navigation receivers*. These units receive and evaluate the dedicated signals transmitted by special ground stations using on-board navigational instruments. The signals are called non-directional beacon (NDB) and use frequencies between 255 kHz and 526.5 kHz, that is, they cover the LW and MW frequency range. There are also the very high frequency omnidirectional radio range (VOR) and the localizer signals for the instrument landing system (ILS) operating in the VHF range. The frequencies are in the range from 108 MHz to 117.975 MHz, which are close to voice radio frequencies. The ILS glide path transmitters operate in the UHF range between 328.6 MHz and 335.4 MHz [8]. VOR systems are often furnished with distance measuring equipment (DME). On request from the on-board instruments they issue pulse pairs that allow the interrogator on board to determine the slant range to the ground station. Internationally, the frequency range from 960 MHz to 1,215 MHz is reserved for this purpose. The importance of so-called markers using 75 MHz is continuously declining. Their beam is directed upwards vertically, so that an aircraft receives the signal only when flying directly over it. They are subdivided into airway markers (at crucial points along air routes like 'intersections', reporting points, and the like) and in ILS markers (indicating the final distances to the landing strip: outer marker at 3.9 NM/7.2 km and middle marker at 3,500 ft/1,050 m). For further information about the systems and services of air traffic navigation please refer to [9], which gives comprehensive coverage of all aviation-related topics.

II.3.2 Maritime Radio

Maritime radio service is one of the earliest fields in which radio communication was used (Fig. II.8). It provides a connection from one ship to another and between ships and coastal stations. It is also used for on-board radio communication. The receivers required for the various modes of communication, both customary or specified, necessitate interference-free processing of all common communication methods. VHF maritime radio does not adhere to consistent channel spacing. Though the frequencies of two successive channels are always 50 kHz apart, some other channels may use frequencies only 25 kHz apart (e.g., channel 60 uses 156.025 MHz and channel 1 uses 156.050 MHz; channel 61 uses 156.075 MHz and channel 2 uses 156.100 MHz). This interlaced channel assignment has historical reasons. Besides the terrestrial mobile maritime radio service, there is also mobile maritime radio service via satellite, like the Inmarsat system, providing voice radio, telex, telefax, and e-mail communications services, or the COSPAS-SARSAT system, conveying nautical distress messages. Broadcasting of safety information is handled by the NAVTEX system via terrestrial MW and SW. Another way for distributing safety information in the form of enhanced group calls uses Immarsat-C.

In addition, there is LORAN-C (long range navigation), the maritime navigation radio system, which is handled aboard ships by the *radio location receiver*. Additional information

Table II.2 Air traffic radio frequency bands according to ITU RR [2]

Band	Range	Frequency (ITU RR region 1)	Frequency (ITU RR region 2)	Frequency (ITU RR region 3)	Emergency frequency
MF	100 m	2,850 kHz–3,155 kHz	2,850 kHz–3,155 kHz	2,850 kHz–3,155 kHz	2,182 kHz
MF/HF	87 m	3,400 kHz–3,500 kHz	3,400 kHz–3,500 kHz	3,400 kHz–3,500 kHz	3,023 kHz
HF	77 m	3,800 kHz–3,950 kHz		3,900 kHz–3,950 kHz	
HF	63 m	4,650 kHz–4,850 kHz	4,650 kHz–4,750 kHz	4,650 kHz–4,750 kHz	5,680 kHz
HF	53 m	5,450 kHz–5,730 kHz	5,450 kHz–5,730 kHz	5,450 kHz–5,730 kHz	
HF	45 m	6,525 kHz–6,765 kHz	6,525 kHz–6,765 kHz	6,525 kHz–6,765 kHz	8,364 kHz
HF	33 m	8,815 kHz–9,040 kHz	8,815 kHz–9,040 kHz	8,815 kHz–9,040 kHz	10.003 MHz
HF	30 m	10.005 MHz–10.1 MHz	10.005 MHz–10.1 MHz	10.005 MHz–10.1 MHz	
HF	27 m	11.175 MHz–11.4 MHz	11.175 MHz–11.4 MHz	11.175 MHz–11.4 MHz	
HF	23 m	13.2 MHz–13.36 MHz	13.2 MHz–13.36 MHz	13.2 MHz–13.36 MHz	14.993 MHz
HF	20 m	15.01 MHz–15.1 MHz	15.01 MHz–15.1 MHz	15.01 MHz–15.1 MHz	
HF	17 m	17.9 MHz–18.03 MHz	17.9 MHz–18.03 MHz	17.9 MHz–18.03 MHz	19.993 MHz
HF	14 m	21.924 MHz–22 MHz	21.924 MHz–22 MHz	21.924 MHz–22 MHz	
HF	13 m	23.2 MHz–23.35 MHz	23.2 MHz–23.35 MHz	23.2 MHz–23.35 MHz	
VHF	2.5 m	117.975 MHz–137 MHz	117.975 MHz–137 MHz	117.975 MHz–137 MHz	121.5 MHz*
VHF	2 m	138 MHz–144 MHz*			
(VHF/UHF	1.3 m–75 cm	225 MHz–400 MHz**)			

* Deviations/limitations according to country may apply.
** Allocated to military air traffic, harmonized throughout Europe.

Figure II.8   Modern VHF transceiver IC-GM651 from ICOM for radio communication in commercial maritime traffic designed for voice radio in class F3E emission with 25 W. During operation on any channel a built-in independent receiver monitors the DSC channel 70 continuously (in class G2B emission). This allows automated emergency communication. Distress alarm messages received can be forwarded to coastal stations. An optional hand-held unit offers comprehensible sound reproduction, even in noisy surroundings. The receiving range covers the range from 156 MHz to 163.425 MHz (and the transmit frequency from 156 MHz to 161.450 MHz). (Company photograph of ICOM.)

about systems and services for radio navigation at sea can be found in [9]. Table II.3 lists marine radio frequencies according to [2].

During the high point of maritime radio on medium and short wave, the (now shut down) radio stations used high quality receivers dedicated to the respective frequency ranges. The term *all-wave receiver* was coined for units that covered as many of the required receiving frequencies bands and classes of emission as possible in one unit. Over time, the meaning of the term changed, however, and now it usually describes a receiver operating on as many frequencies as possible, allowing demodulation of various emission classes (Fig. II.9).

II.3.3   Land Radio

Unlike aeronautical radio and maritime radio the term land radio does not characterize a certain group of users or the specific frequency ranges they use. According to [2] and [7] land radio users include individual licence holders with stationary or mobile radio equipment who are not members of an organized radio service. The range of users is so wide that only some examples are listed below.

- Radio receivers typically used by embassies, news agencies or press services using short wave transmission are called radiotelephony terminals (Figs. II.9 till II.11). Generally, these *station receivers* are designed for uninterrupted receive mode, often using specific frequencies or channels to generate receive protocols at any desired time. (They are not always equipped with a flywheel control for manual frequency tuning, but may have numerical keys for direct frequency selection.) As a rule, such receivers meet very high quality standards in respect to the receiver characteristics (Part III). All-wave receivers as described in Section II.3.2 are used for many such applications.

Table II.3 Marine radio frequency bands according to ITU RR [2]

Band	Range	Frequency (ITU RR region 1)	Frequency (ITU RR region 2)	Frequency (ITU RR region 3)	Emergency/calling frequency	DSC frequency	Telex frequency
VLF	21.4–15 km	14 kHz–19.95 kHz*	14 kHz–19.95 kHz*	14 kHz–19.95 kHz*			
VLF/LF	15 km–4.3 km	20.05 kHz–70 kHz*	20.05 kHz–70 kHz*	20.05 kHz–70 kHz*			
LF	4 km	72 kHz–84 kHz	70 kHz–90 kHz	70 kHz–90 kHz			
LF	3.5 km	86 kHz–90 kHz					
LF	2.7 km	110 kHz–112 kHz	110 kHz–160 kHz	110 kHz–160 kHz*			
LF	2.5 km	115 kHz–126 kHz*					
LF	2.2 km	129 kHz–148.5 kHz*	129 kHz–148.5 kHz*				
MF	720 m–570 m	415 kHz–526.5 kHz*	415 kHz–510 kHz	415 kHz–526.5 kHz	500 kHz		
MF	175 m	1,606.5 kHz–1,800 kHz					
MF	140 m	2,045 kHz–2,160 kHz	2,065 kHz–2,107 kHz*	2,065 kHz–2,107 kHz*			
MF	137 m	2,170 kHz–2,194 kHz	2,170 kHz–2,194 kHz	2,170 kHz–2,194 kHz	2,182 kHz	2,187.5 kHz	2,175.5 kHz
MF	115 m	2,625 kHz–2,650 kHz					
HF	75 m	4,000 kHz–4,438 kHz	4,000 kHz–4,438 kHz	4,000 kHz–4,438 kHz	4,125 kHz	4,207.5 kHz	4,177.5 kHz*
HF	45 m	6,200 kHz–6,525 kHz	6,200 kHz–6,525 kHz	6,200 kHz–6,525 kHz	6,215 kHz	6,312 kHz	6,268 kHz
HF	35 m	8,100 kHz–8,815 kHz	8,100 kHz–8,815 kHz	8,100 kHz–8,815 kHz	8,291 kHz	8,414.5 kHz	8,376.5 kHz
HF	25 m	12.23 MHz–13.2 MHz	12.23 MHz–13.2 MHz	12.23 MHz–13.2 MHz	12.29 MHz	12.577 MHz	12.52 MHz
HF	18 m	16.36 MHz–17.41 MHz	16.36 MHz–17.41 MHz	16.36 MHz–17.41 MHz	16.42 MHz	16.804,5 MHz	16.695 MHz
HF	16 m	18.78 MHz–18.9 MHz	18.78 MHz–18.9 MHz	18.78 MHz–18.9 MHz			
HF	15 m	19.68 MHz–19.8 MHz	19.68 MHz–19.8 MHz	19.68 MHz–19.8 MHz			
HF	14 m	22 MHz–22.855 MHz	22 MHz–22.855 MHz	22 MHz–22.855 MHz			
HF	12 m	25.07 MHz–25.21 MHz	25.07 MHz–25.21 MHz	25.07 MHz–25.21 MHz			
HF	11 m	26.1 MHz–26.175 MHz	26.1 MHz–26.175 MHz	26.1 MHz–26.175 MHz			
VHF	2 m	138 MHz–144 MHz*					
VHF	2 m	156 MHz–162.05 MHz	156 MHz–162.05 MHz	156 MHz–162.05 MHz	156.8 MHz (Ch 16)	156.525 MHz (Ch 70)*	
VHF	1.4 m		216 MHz–220 MHz				

*Deviations/limitations according to country may apply.

Figure II.9 Front view of a typical sturdy VLF/HF receiver meeting the highest demands in regard to both receiver characteristics (Part III) and electromechanical properties. The photo shows the RA6790/GM from RACAL, designed as a multi-conversion superhet (Section I.2.2) with fully analog signal processing. The easily rotating flywheel control to the right of the centre is for manual station search. The unit can be equipped with up to seven different IF filters and, at the same time, functions as an all-wave receiver featuring a receiving frequency range from 500 Hz to 30 MHz for AM, CW, FM, ISB (optional) and SSB demodulation. This unit is a 3 HE 19″ plug-in module.

- In the public and private commercial telephony segment (named professional mobile radio (PMR) users such as government agencies and organizations performing security assignments, power corporations, taxi radio, and transport services, communicate on assigned frequency channels (Fig. II.11). The receiver section of the transceivers, designed as walkie-talkies or mobile radio equipment, are preprogrammed to the assigned channels. The operating frequency range for such units often cover an entire band for one class of emission. In fact, the respective programming of the firmware allows equipment operation only in the channels assigned to the relevant organization. After simple reprogramming in compliance with the frequency assignment stipulations of the authorities the same product can be sold to a different user.

II.3.4 Amateur Radio

The receiver sections of transceivers typically used by amateur radio service (Fig. II.12) are designed as either:

- monoband or multiband units for demodulation of a certain class of emission,

or

- *all-wave receivers* as described in Section II.3.2. They typically enable the demodulation of emission classes A1A, A3E, F3E, and J3E (upper and lower sideband). For the demodulation of digital emission classes like F2D, here often audio frequency shift keying (AFSK), most units require an additional decoder connected to the AF output. High-end equipment often allows selection of the receive bandwidth (Section III.6.1). These units are usually designed and optimized for manual search mode and feature a flywheel control for frequency tuning. *Amateur receiver* is another commonly used name.

The equipment is designed for tuning across the receiving frequency range (Table II.4) either continuously or in small frequency increments. In fact, there is no standardized

Figure II.10 Equipment family of the M3SR 4100 series from Rohde&Schwarz includes the model EK4100, a state-of-the-art receiver. Also available are transceivers with 150 W, 500 W or 1 kW HF transmit power (the rack in the background shows the 1 kW HF transmit stage and the required power supply unit). The units form a platform [10] that can be flexibly configured and upgraded via software. They are used by various classical conventional radio services as well as by the military. (When communicating on short waves one does not rely on telephone networks or cellular radio. HF communication equipment is comparatively easy to install and operates reliably. This applies particularly to areas of weak infrastructure or in emergency or natural disaster scenarios.)

The receive frequency range is from 10 kHz to 30 MHz, with AM, CW, FM, ISB, and SSB demodulation and several data transfer formats. Dimensions: (receiver) 3 HE 19″ plug-in module. (Company photograph of Rohde&Schwarz.)

channel spacing. Owing to the fact that the class of emission used covers a certain bandwith the term channel is used when referring to the transmit frequency range. (When operating via amateur radio satellites, the term uncoordinated multiple access to the satellite is used.)

Licensed radio amateurs' applications require optimum receiver characteristics, such as in the uses described in Sections II.4 and II.5 and in the ocean navigation applications described in Section II.3.2, as well as in certain other fields of use. Also in amateur radio

Figure II.11 Modern transceiver (radio equipment) TELCOR R125F from Telefunken Radio Communication Systems for wireless wide-area communication designed for voice radio, telefax and data traffic in the emission class J3E with 125 W. Within the entire received frequency range from 0.1 MHz to 30 MHz (1.6 MHz to 30 MHz transmission) up to 200 communication channels can be programmed in compliance with channel assignments. As an option, a GPS receiver can be integrated in the housing, enabling the operator to determine and transmit his own location without problems. (Company photograph of Telefunken Radio Communication Systems.)

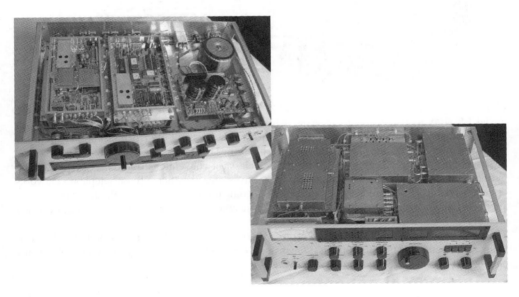

Figure II.12 The DIY receiver shown was built by a licensed radio amateur and represents a complete short-wave radio [11]. It is a single-conversion superhet (Section I.2.1) with an intermediate frequency of 9 MHz. The consistent use of large-sized ferrite cores in the HF preselector and the bolting of the surfaces for ground connection to the lid and bottom plate guarantees a 100 dB far-off selection (Section III.11) with the preselector. A MOSFET switching mixer with diplexer for terminating the mixer results in excellent large-signal behaviour (Section III.12) and interference-free reception of AM, CW, and SSB modes with a noise figure of only 12 dB (Section III.4.2).

Table II.4 Amateur radio frequency bands according to ITU RR [2]

Band	Range	Frequency (ITU RR region 1)	Frequency (ITU RR region 2)	Frequency (ITU RR region 3)
LF	2,200 m	135.7 kHz–137.8 kHz*	135.7 kHz–137.8 kHz*	135.7 kHz–137.8 kHz*
MF	630 m	472 kHz–479 kHz*	472 kHz–479 kHz	472 kHz–479 kHz*
MF	160 m	1,750 kHz–2,000 kHz*	1,800 kHz–2,000 kHz	1,800 kHz–2,000 kHz*
HF	80 m	3,500 kHz–3,800 kHz	3,500 kHz–4,000 kHz*	3,500 kHz–3,900 kHz
(HF	56 m	5,250 kHz–5,450 kHz*	5,250 kHz–5,450 kHz*	5,250 kHz–5,450 kHz*)
HF	40 m	7,000 kHz–7,200 kHz*	7,000 kHz–7,300 kHz*	7,000 kHz–7,200 kHz*
HF	30 m	10.1 MHz–10.15 MHz	10.1 MHz–10.15 MHz	10.1 MHz–10.15 MHz
HF	20 m	14 MHz–14.35 MHz	14 MHz–14.35 MHz	14 MHz–14.35 MHz
HF	17 m	18.068 MHz–18.168 MHz	18.068 MHz–18.168 MHz	18.068 MHz–18.168 MHz
HF	15 m	21 MHz–21.45 MHz	21 MHz–21.45 MHz	21 MHz–21.45 MHz
HF	12 m	24.89 MHz–24.99 MHz	24.89 MHz–24.99 MHz	24.89 MHz–24.99 MHz
HF	10 m	28 MHz–29.7 MHz	28 MHz–29.7 MHz	28 MHz–29.7 MHz
VHF	6 m	50 MHz–52 MHz*	50 MHz–54 MHz*	50 MHz–54 MHz*
(VHF	4 m	70 MHz–70.5 MHz*)		
VHF	2 m	144 MHz–146 MHz	144 MHz–148 MHz	144 MHz–148 MHz
VHF	1.3 m		220 MHz–225 MHz	
UHF	70 cm	430 MHz–440 MHz*	430 MHz–440 MHz*/**	430 MHz–440 MHz*/**
UHF	33 cm		902 MHz–928 MHz	
UHF	23 cm	1.24 GHz–1.3 GHz	1.24 GHz–1.3 GHz	1.24 GHz–1.3 GHz
UHF	13 cm	2.3 GHz–2.45 GHz	2.3 GHz–2.45 GHz	2.3 GHz–2.45 GHz
UHF	9 cm	3.4 GHz–3.475 GHz*	3.3 GHz–3.5 GHz	3.3 GHz–3.5 GHz
SHF	6 cm	5.65 GHz–5.85 GHz	5.65 GHz–5.925 GHz	5.65 GHz–5.85 GHz
SHF	3 cm	10 GHz–10.5 GHz	10 GHz–10.5 GHz	10 GHz–10.5 GHz
SHF	1.2 cm	24 GHz–24.25 GHz	24 GHz–24.25 GHz	24 GHz–24.25 GHz
EHF	6 mm	47 GHz–47.2 GHz	47 GHz–47.2 GHz	47 GHz–47.2 GHz
EHF	4 mm	76 GHz–81 GHz	76 GHz–81 GHz	76 GHz–81 GHz
EHF	2.5 mm	122.25 GHz–123 GHz	122.25 GHz–123 GHz	122.25 GHz–123 GHz
EHF	2 mm	134 GHz–141 GHz	134 GHz–141 GHz	134 GHz–141 GHz
EHF	1.2 mm	241 GHz–250 GHz	241 GHz–250 GHz	241 GHz–250 GHz

*Deviations/limitations according to country may apply.
**In Australia, USA, Jamaica and the Philippines, the bands 420–430 MHz and 440–450 MHz are additionally allocated to the amateur service on a secondary basis.

services one may find the most sophisticated receiver designs, produced in relatively large numbers. This is especially true for processing transmitted signals of emission classes A1A, A3E, F3E, and J3E and for the large-signal immunity (Section III.12).

II.3.5 Mobile Radio

Today, the term 'mobile radio' usually describes the public mobile radio network for telephony using mobile radio stations to communicate with each other or with terminals of the public landline network. The expression 'land mobile radio service' originated at a time when mobile carphone with analog frequency modulation was used. It only insufficiently characterizes the present and future possibilities of mobile radio technology. Of course there are also satellite-supported mobile networks, which are used especially in areas where the terrestrial mobile (or cellphone) networks provide no or merely insufficient coverage.

The receiver of a portable telephone (cell phone) must meet stringent requirements in terms of the applicable standards (Section III.2.2). This subject is covered in detail in the relevant literature. For further information please refer to [1], [12] and [13].

II.4 Radio Intelligence, Radio Surveillance

The objective of radio intelligence and radio surveillance (also referred to as radio monitoring, signal intelligence, or radio interception) is to obtain a complete picture of current radio operation over a wide frequency range. Dedicated equipment allows us to establish when, from where, and by whom a certain frequency is occupied within the frequency range monitored (Table II.5 explains some specific terms). Furthermore, the news content of the selected emission of interest, determined by searching the frequency band, is gathered and evaluated [14].

According to [15], radio intelligence systems consist of several units like a search receiver, surveillance receiver, hand-off receiver, measuring receiver/analysis receiver, operating stations and direction finders (DF) (Section II.5.2) in combination with direction-finding receivers, where applicable. (Explanations of the single components and information on their fields of use will be given in the sections below.) These may be combined in various configurations according to the task at hand. Usually the individual receivers of such systems are coupled to form functional groups. The functional groups contain a large number of elements (assemblies) and are increasingly included in multi-functional systems by integrating the discrete elements required (Fig. II.13). Such concepts allow accessing all receivers of a functional group from one common user interface with display (see Fig. II.26). The single receiver modules form a 'window' in the user interface [15]. Usually they are linked via a Gigabit LAN with high-performance PC clusters in the background. This enables very flexible adaptation to meet the operational requirements of the relevant job. At the start of a work session they may be used for rapid search functions and can later be conditioned for surveillance, analysis or hand-off reception.

An increasing number of multi-function stand-alone units capable of coping with the various tasks described in Sections II.4.2 to II.4.4 are commonly referred to as *monitoring receivers* (Figs. II.13 and II.14). However, the name monitoring receiver is also used for *control* (measuring) *receivers* used to continuously monitor emissions (e.g., directly at public broadcasting stations) to check the compliance of their transmit quality.

II.4.1 Numerous Signal Types

In classic radio systems each user is normally provided with a separate radio channel which remains continuously occupied by the emission during the active radio communication. In order to serve several users the radio channels are directly adjacent to each other within the frequency range (Fig. II.15). This is the classical frequency division multiple access

Table II.5 Explanation of specific terms

Term	Explanation
Bearing basis	Geographical arrangement of several DF stations or antennas for determining the origin of electromagnetic radiation sources
Detection probability	Probability of detecting and identifying (classifying) a signal
Emission	Deliberate radiation of electromagnetic energy
Evaluation/analysis	Processing of detected results with the objective of interpretation and utilization
Monitoring probability/ interception probability	Probability of intercepting an electromagnetic radiation, depending on the performance data of the monitoring receiver
Radiation	Energy flow in the form of electromagnetic waves emitted from any source
Radio bearing/radio direction finding	Determining the direction of incident electromagnetic waves by suitable means; the *bearing* is the final result
Radio intelligence	Collective term for measures used to obtain information on the existing frequency spectrum for analysis and evaluation (the term radio intelligence is predominantly used in military and intelligence services)
Radio location	Determining the location, speed and/or other properties of an object, or obtaining information on these parameters on the basis of the propagation characteristics of electromagnetic waves
Radio monitoring	Monitoring and evaluating radio communications to ensure proper performance
Radio surveillance	Searching for and/or picking up (and determining the direction of) an emission
Signal classification	Detecting the type of modulation, the coding technique used, and other decisive parameters of a detected or monitored emission as well as its classification based on these properties
Single station locator (SSL)	Bearing system allowing the determination of the position of a transmitter from only one location

Source: Freely adapted from the ITU Radio Regulations

mode (FDMA). Several of today's communication systems emit their signals not only on or around a constant centre frequency, but use time-variant techniques and utilize several frequencies. Examples of wideband signals are:

- Frequency-agile emissions (also called chirp, sweep, or stepping emissions) are used, for example, in various radar applications and ionospheric research.
- Frequency hopping is used, for example, for cordless telephones and in several critical wireless military applications.

Figure II.13 ESMD compact wideband monitoring receiver from Rohde&Schwarz. It forms the centrepiece of a modern radio intelligence station comprising a search receiver, analysis receiver and a wide range of multi-channel demodulation capabilities. The unit can be upgraded for single-channel direction-finding. Fully equipped, it covers a receive frequency range from 9 kHz to 26.5 GHz. Performance data include: Maximum real-time bandwidth of 80 MHz, spectral image build-up with gap-free dynamically overlapping FFT; AM, CW, FM, PM, ISB, SSB, PULSE, analog TV, and IQ demodulation. The dimensions are 426 × 176 × 450 mm. (Company photograph of Rohde&Schwarz.)

- Spread-spectrum emissions are used, for example, in the global positioning system (GPS) or in wireless local area networks (WLAN).
- Cellular systems on the basis of multiple-access methods (also called frequency/time/code division multiple access) are used, for example, in different digital (mobile) radio standards.

It is also possible to change the frequency in accordance with a pattern known only to the communication partners. Using these LPI (low probability of intercept) or LPD (low

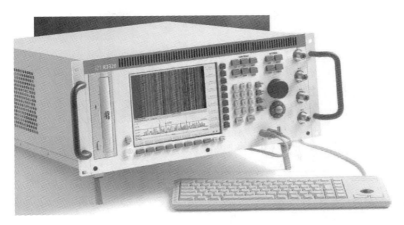

Figure II.14 R3320 modern short-wave surveillance receiver, from the Fraunhofer (research organization) spin-off Innovationszentrum für Telekommunikationstechnik. It functions as a digital receiver of generation 2.5 (Fig. I.19), i.e. quasi fully digital. Besides the real-time bandwidth of up to 24 MHz it allows FFT-based spectrum analysis with a frequency resolution of better than 2 Hz and the simultaneous demodulation of several received frequencies. Performance data include: Receive frequency range 9 kHz to 32 MHz, maximum real-time bandwidth 24 MHz (deactivated preselector, Section III.11), spectral image build-up with up to 1,000 FFTs per second, FM demodulation and IQ offline analysis, 19″ plug-in module of 3HE. (Company photograph of Innovationszentrum für Telekommunikationstechnik.)

probability of detection) signals is an effective method for thwarting attempts to detect the transmitter, intercept information, or disturb the emission.

Figure II.16 illustrates the basic patterns of signal types in current usage [16].

With emissions in *chirp mode*, the carrier frequency changes almost continuously.

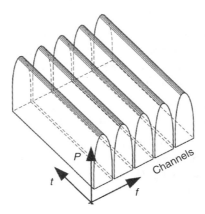

Figure II.15 In FDMA the channel used or assigned to the respective user is available with the full signal-to-interference ratio for the entire duration of the emission. In order to serve several users, the radio channels (with different activities) are adjacent to each other within the frequency range.

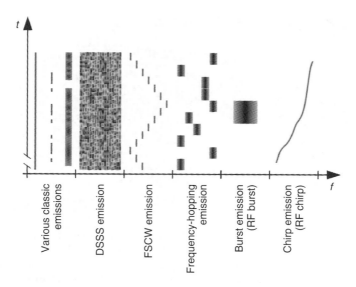

Figure II.16 Illustration of the principle of time and frequency patterns of various frequency-agile emissions, complemented by classical emission types using constant centre frequencies (fixed-frequency emissions).

With emissions in *frequency-stepped continuous wave mode* (FSCW) the carrier frequency is varied in short time intervals by a known increment. The simulated frequency ramp often serves in various radar applications to accomplish tasks like distance measurements.

With emissions in *frequency hopping mode* the message is split into short segments (typically a few ms long, depending on the transmit frequency range) and transmitted with varying carrier frequencies. (With LPI signals in security-relevant applications these are often pseudo-random variations.) The targeted receiver follows this prearranged (pseudo-random) sequence to recover the information contents of the transmission.

With emissions in *direct sequence spread spectrum mode* (DSSS) the signal is subjected additionally to a pseudo-random phase modulation in order to distribute the useful signal over a wide frequency band. The noise-like character and the low power density of the spread signal make detection very difficult.

With emissions in *short-term transmit mode* the information to be conveyed is compressed. Emissions takes place in a short powerful broadband burst [16].

With emissions in *time division multiplexing mode* (also called time division multiple access (TDMA)) the participant can use the entire system bandwidth and the full interference ratio for a brief time segment. The time segment assigned returns within the frame clock time (Fig. II.17). This means that with time division multiple access (TDMA) several participants share one channel. (Such networks are well suited for communicating digital information. When a TDMA network is used for voice transmission, the data are compressed in time and transmitted within the time slot assigned. The timeslots are periodically repeated according to the cycle frequency used. Such a cycle period is correctly called a frame. The principle is illustrated in Figure II.76.)

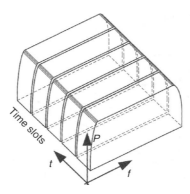

Figure II.17 In TDMA the individual user has access to the entire system bandwidth with the full interference ratio, but in a brief time slot only. The time slot method allows serving several participants.

With emissions in *code multiplex mode* the transmit channel is not restricted either in time or in frequency, but the power density is limited. Even with all users active, the interference effect on the individual authorized participant must still be tolerable (Fig. II.18). The message is coded on the transmitter side by a code word. Signal retrieval on the receiver side is accomplished by synchronization with the same code word. If this is not known, gathering of the information contents is not possible [17]. With code division multiple access (CDMA) the users (e.g., 128 participants) share one wideband transmit channel. This channel is used simultaneously for both transmitting and receiving, and the participants are separated by 'de-spreading codes' [18].

II.4.2 Searching and Detecting

A common method for detecting emissions is to search the frequency range of interest by means of a *search receiver* offering an automatic (or manual) mode for rapid tuning. These receivers are often designed as *wideband receivers*. They are used in signal detection or

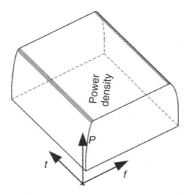

Figure II.18 In CDMA all users have continuous access to the entire width of the frequency band. These are separated by 'de-spreading' codes.

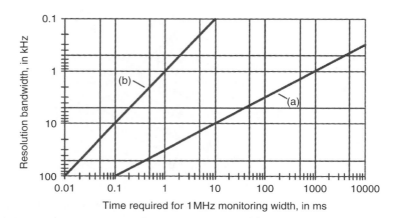

Time required for 1 MHz monitoring width, in ms

Figure II.19 Minimum time required for searching sections of a frequency range simultaneously using FFT (b) or sequential mode (a), depending on the resolution bandwidth [19]. In this example the frequency window is 1 MHz. As can be seen, the time required becomes shorter with decreasing resolution or selection. This suggests that the scan mode is particularly interesting in the VHF/UHF range, where most tasks do not require narrow-band selection.

signal acquisition and are therefore also known as *acquisition receivers*. In military or intelligence-related reconnaissance applications they may be referred to as *reconnaissance receivers*.

All of the signals detected in the frequency range investigated can be presented on a display as a spectral panorama. Basically, there are two known methods used by these receivers:

(a) *Sequential search* for rapid scanning of the interesting frequency range by means of a wobble injection oscillator (Fig. I.4) in the search receiver (*scanner receiver*). With a constant channel spacing or frequency pattern the receive frequency is searched by stepping from channel to channel. When this is not possible, continuous observation is achieved by retuning the receiver in increments of the receive bandwidth used (Section III.6.1). Depending on the scanning rate (Section III.20) there are always some time windows through which only a small segment of the frequency range of interest (with the width of the receive bandwidth used) can be observed. The search rate is ideally proportional to the square of the receive bandwidth which, in turn, depends on the required resolution (Fig. II.19). This situation can be improved by using several receivers in parallel, which deliver their results simultaneously with every step throughout the frequency range.

(b) *Simultaneous search range for range* with so-called FFT *multi-channel receivers* using the fast Fourier transformation (FFT) on a group of 'quasi receive channels' arranged in parallel. This allows the quasi-parallel detection of all signal activities within a certain frequency segment virtually in real-time. (In the beginning this equipment was called *filter bank receiver*.) First, the respective segment of the frequency spectrum of interest was mixed to a mostly wideband IF layer and then divided into a large

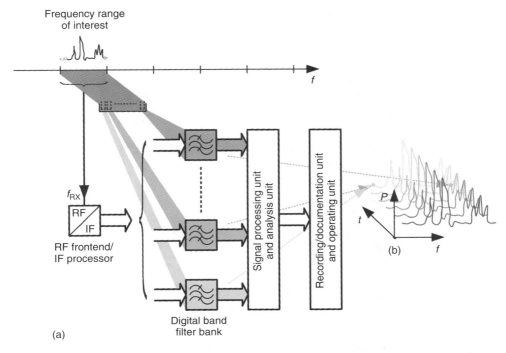

Figure II.20 Principle of the simultaneous multi-channel search for signal detection using an FFT multi-channel receiver. A wide frequency range is split up to form several individual channels which are subjected to synchronous processing and the results are then displayed in either a waterfall diagram or a histogram (Fig. II.21). In this way it is possible to monitor a wider frequency segment simultaneously and continuously.

number of neighbouring channels of equal bandwidths or lines derived from these (Fig. II.20). Finally, the signals were processed simultaneously [20]. The width of the simultaneously observed frequency segments is called the real-time bandwidth. The temporal resolution achieved with FFT analysis is the reciprocal value of the frequency width [21]. The search rate is inversely proportional to the resolution bandwidth (Equation (I.9)) of the generated channels or lines (Fig. II.19).

II.4.2.1   Problems with Frequency-Agile Signals and LPI Emissions

Wideband and/or variable-frequency emissions such as those described in Section II.4.1 are hardly ever detected by receivers with discrete tuning. Furthermore, in many intelligence-related applications neither the type nor the pattern of the varying signal parameters is usually known. (Frequency hopping can be detected at least in part by demodulating conventional narrow-band signals. Since the dwell time on one frequency is relatively long and the energy in this channel must be clearly above the noise threshold

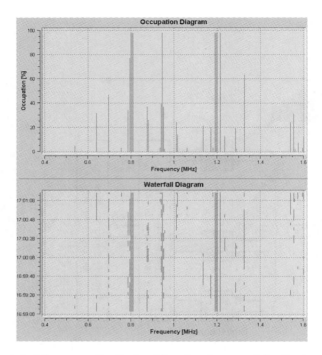

Figure II.21 Statistical frequency distribution (above) and temporal distribution (below) displayed as a waterfall diagram over the monitored period with 1.2 MHz real-time bandwidth. It is apparent that two transmitters with 0.8 MHz and 1.2 MHz are continuously active (these are the medium-wave broadcasting stations of the Bavarian Radio in Ismaning on 801 kHz and Voice of America on 1.197 kHz. (Company photograph of Rohde&Schwarz.)

(compared with spread-band transmissions) a crackling noise or hum may be heard. Such sounds occur when frequency hops coincide with the tuned receive frequency, that is, they are within the received bandwidth. Chirp emissions may also be recognized by a short transient tone. This occurs at the moment when the chirp sounder passes the tuned-in receive frequency and wanders over the receive bandwidth. Spread-band transmissions only cause an increase in the noise level.)

The *detection probability* can be improved with an *automated sequential search* by repeatedly scanning the frequency range of interest (scanning method). The probability of detecting frequency-hopping signals becomes higher the faster the search run (Section III.20). For physical reasons it is an essential requirement that the dwell time on the receive bandwidth of the scanner receiver is longer than or equal to the settling time. This transient settling time is given by the reciprocal value of the receive bandwidth according to

$$t_{\text{dwell FH}}(B_{-6\,\text{dB}}) \geq \frac{1}{B_{-6\,\text{dB}}} \tag{II.1}$$

where

$t_{\text{dwell FH}}(B_{-6\,\text{dB}})$ = required dwell time of the frequency-hopping signal at the receive
$\qquad\qquad$ bandwidth $(B_{-6\,\text{dB}})$ used with the scanning method, in s
$\quad B_{-6\,\text{dB}}$ = bandwidth (−6 dB bandwidth) of the receiver, in Hz

With a selected receive bandwidth of, for example, 10 kHz this means that the dwell time of the hop on the receive frequency must be

$$t_{\text{dwell FH}}(B_{-6\,\text{dB}} = 10\,\text{kHz}) \geq \frac{1}{10\,\text{kHz}} \geq 100\,\mu s$$

The rate of varying the frequency hopping must therefore be below 100 MHz/s. Since the probability of detection depends on the ratio between the short dwell time and the total duration of the search run, a single frequency jump can be detected with some certainty only if the dwell time (on one frequency) is longer than the search run or, in other words, if all frequencies of interest are examined within the dwell time. Otherwise the search must be repeated several times. The advantage of this method is that relatively few equipment components are needed [19].

An extremely high success rate is obtained when using the *simultaneous search range-for-range*. If the FFT multi-channel receiver offers a real-time bandwidth sufficiently wide to observe the entire frequency range covered by the frequency-jump signal, even a single frequency jump would be detected with a probability of 100%. For reconnaissance and surveillance applications as wide a continuous frequency band as possible is always favourable for FFT analysis (Fig. II.22). However, the disadvantage of using wide frequency bands is that the computing power required increases disproportionately with the number of FFT lines (corresponding to the frequency range). The calculation complexity for n FFT lines is

$$O\left(n \cdot \log(n)\right) \tag{II.2}$$

where
$O =$ complexity of FFT, dimensionless
$n =$ number of FFT lines, dimensionless

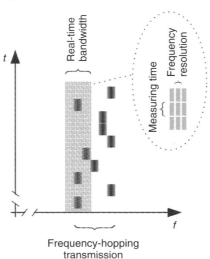

Figure II.22 Wideband detection for a case in which the search width is restricted to a real-time bandwidth too narrow for detecting the wideband signal (using the frequency-hopping signal of Figure II.16 as an example). Only some of the hops can be detected, but these are detected with full temporal resolution.

In addition, the dynamics of such an arrangement decreases with increasing frequency width. Error signals or ghost signals occur more often with wider frequency bands, partly due to intermodulation (Section III.9). This clearly illustrates the problems existing with a continuous surveillance having larger frequency ranges [21]. The real-time bandwidth is therefore limited due to the sampling rate, the dynamic range of the A/D converter (Section I.3.2), and the processing speed required. In order to still maximize the detection probability with a wider frequency band, current systems restart the scanning operation after tuning to another frequency spaced at a distance equal to the real-time bandwidth in order to cover the next surveillance window [16]. (The FFT analysis is repeated in short intervals covering another limited width.) However, there are always surveillance time gaps (Fig. II.23) in which single hops of frequency hopping signals are (or can be) lost. It is clear that the demands of high temporal and high frequency resolutions are in conflict with each other. It is therefore suggested that at least one additional broadband analysis channel be used that is synchronized with the other channels [21]. The entire frequency spectrum with high temporal resolution would then be available continuously in this analysis channel for surveillance and analysis purposes (Fig. II.24). The bandwidth of the analysis channel should preferably be the same as that of the frequency band to be monitored. This enables coverage of the entire frequency band to be monitored with a high temporal resolution while obtaining a high frequency resolution by applying a selection with high dynamics in the other channels of relatively narrow bandwidth. In order to counteract any errors resulting from the large bandwidth of the analysis channel,

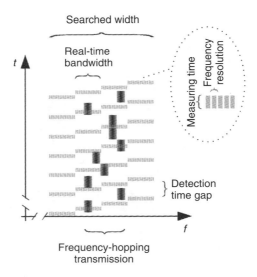

Figure II.23   Wideband detection for a case in which the search width is a multiple of the real-time bandwidth (using the frequency-hopping signal from Figure II.16 as an example). If the acquisition receiver in use allows the search width to be adapted flexibly to the task, the example shown would provide a maximum detection rate by reducing the search width to twice the real-time bandwidth, since the time gaps would be drastically reduced.

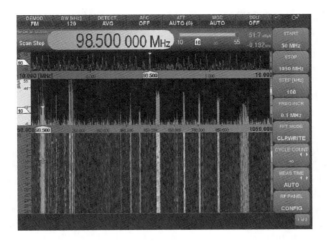

Figure II.24 Panorama scan from 50 MHz to 1,050 MHz using the Rohde&Schwarz ESMD receiver (Fig. II.13). The spacing between the FFT lines is 100 kHz. With this setting the span of 1 GHz is scanned 20 times per second. (But despite the high scanning rate there is still a time gap of 50 ms until the same FFT line is scanned again.) The VHF FM broadcast band can be seen at the left margin of the spectrum, while at the right margin the GSM band is visible at 900 MHz. (Company photograph of Rohde&Schwarz.)

the evaluation is carried out only after a signal occurs in both the analysis channel and at least one other channel [21].

II.4.3 Monitoring Emissions

The data acquired are used for presenting and evaluating the activity and preparing the *hand-off receiver*. These receivers are intended for evaluating the information content of selected signals. They are tuned to the frequency of the selected signal or transmitting station, and operate independently of the search mode. These receivers can often be controlled with a movable cursor. When automatically analyzing or listening to the signal the operator can decide whether or not a certain emission is of importance for the current surveillance task, that is, whether or not it belongs to a certain radio network. If the transmission giving rise to the detection of signals ends, the frequency adjusted at the hand-off receiver can be used for on-going surveillance [22].

In the same way, [14] describes the so-called *query receiver*, which jumps back and forth between a few receive frequencies that are actually of interest. Within a relatively short period it queries all those frequencies classified by the user as being of interest. If any activity is detected in one of the channels monitored, the device will automatically activate a hand-off receiver for continued surveillance.

A large number of narrow-band emissions may be contained in the wider frequency range covered by the search receiver. These often follow a certain channel pattern. When

Figure II.25 IC-R8500 semi-professional broadband communications receiver from ICOM optimized for manually varying the tuned-in frequency. A recorder for registering the signals received
can be connected to the REC or REC-Remote sockets on the front panel. The unit allows different
query modes like skip memory, auto-storage and selective emission class search run. Performance
data include: receive frequency range from 100 kHz to 2 GHz (>1 GHz with limitations); memory
search run on up to 1000 programmed channels with a moderate search speed of 40 channels/s; AM,
CW, FM, and SSB demodulation. The dimensions are 287 × 112 × 309 mm. (Company photograph
of ICOM.)

interested in the information contents of the narrow-band emission for analysis purposes,
it should be considered that hand-off receivers are generally singe-channel units that are
typically capable of demodulating one signal only. Several communications receivers
are often used in parallel to provide the possibility of demodulating signals of different
emission classes and with different receive bandwidths (Section III.6.1). These are similar
to the all-wave receivers described in Sections II.3.2 and II.3.4 (Fig. II.26). They are
capable of processing emission classes A1A (CW), A3E (AM), B8E (ISB), F3E (FM),
G3E (PM), J3E (SSB, upper and lower sideband). Newer models also allow complex
IQ demodulation, which can be used to make other emission classes and radio standards
accessible by using software demodulators or suitable analysis software.

The parallel reception of GMS signals will be used as an example for demonstrating
the problematic nature of the multi-channel demodulation necessary. When acquiring
part of the GSM band at 945 MHz with a receive bandwidth of 9.6 MHz, this segment
contains all emissions originating from the respective GMS downlinks in the 200 kHz
channel pattern active at the time. Though modern receivers have a real-time bandwidth
enabling them to receive up to 48 contained downlink signals, they can demodulate either
one channel only in a narrow band of 200 kHz or the entire non-separated mix having
a bandwidth of 9.6 MHz ('single-channel mode'). To compensate for the disadvantage
of 'single-channel mode' [23] suggests the implementation of a digital selection filter
bank with demodulator to operate within a larger receive bandwidth and to demodulate
independently and simultaneously up to 8,192 channels with a reasonable amount of
hardware. As with parallel receivers, the demodulation mode can be selected for each of
the channels independently. This is possible with the combination of a DFT polyphase
filter bank and a demodulator functioning in time slot mode (time multiplexing technique).
The effective computational effort is comparable to that of a single selection filter and
a single demodulator plus subsequent fast Fourier transformation. The only additional
computation needed, compared with a single-channel digital receiver (Section I.2.4), is
the FFT calculation of a length corresponding to the number of signals to be demodulated.

Figure II.26 EM010 receiver module on VXI basis from Rohde&Schwarz used as a functional element in a modern surveillance system. It may be used for hand-off reception or for technical analysis in the AMMOS radio intelligence system from the same manufacturer. It can be operated in parallel with several other receive paths. Performance data include: receive frequency range from 10 kHz to 30 MHz (optionally 300 Hz to 60 kHz via another input); memory search run on up to 1,000 programmed channels; AM, CW, FM, ISB, and SSB demodulation. (Company photograph of Rohde&Schwarz.)

One of the essential signal processing operations of a digital receiver is the sample rate reduction (Fig. I.24). It utilizes a major portion of the computing power required for digital signal processing, as demonstrated in the following example. A digital receiver uses an A/D converter with a sampling rate of 76.8 MHz. But the reproduction rate (number of sampled values) of the digital AF output is only 32 kHz. In order to convert these sampling rates without aliasing (Section I.3.7) several decimation filters are necessary. Decimating by the factor M means that only one sampling value remains after M sample values pass the decimation filter. All but $M - 1$ sampling values are discarded during the down-sampling process. In the above example, this is repeated several times until 32 kS/s have accumulated. The decision as to which of the M sampling values are retained during decimation is analogous to the decision as to which of the polyphases 0 to $M - 1$ of the filtered signal are to be retained. The combination of filtering and down-sampling can be described by means of a poly-phase decimation filter (Fig. II.27). We then arrive at the general structure of the poly-phase decimator with M branches and a clock reduction by factor M. For the GSM example described this means that an FM demodulator is capable of processing many decimated channels in time-multiplex mode, provide that it has the calculating power for processing the full receive bandwidth. Thus, it is possible to demodulate each of the GSM downlinks independently with one and the same demodulator. Instead of calculating one spectrum only (like with the FFT multi-channel receivers

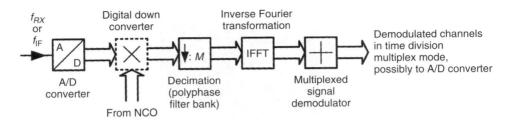

Figure II.27   Functional blocks behind the A/D converter of a digital multi-channel receiver with a polyphase analysis bank for the collective independent demodulation of several channels received together. After the inverse discrete Fourier transformation (IDFT or IFFT) the sampling values of the individual channels are available in time multiplex mode. The demodulator uses the time slot method, allowing the selection of the desired demodulation mode individually for each channel. The mixer enables adjusting the position of the channel pattern and tuning the wideband receive frequency band. Thus, the mixer is not an essential component [23].

described in Section II.4.2) this method additionally provides a selected baseband, a demodulated signal (e.g., in A3E, F3E or complex IQ demodulation), and the required measuring values for each frequency contained in the spectrum. The combination of these techniques results in many virtually independent receivers [23].

II.4.4   Classifying and Analyzing Radio Scenarios

Depending on the parameters selected, certain applications require filtering and/or sorting the results obtained due to the large amount of data that may be extended to include the direction of incidence determined (Section II.5). The main objective is the detection or recognition of signals, transmitting stations, and methods to obtain the respective (tactical) information. Automatic process recognition (Fig. II.28), which can also be used for the continuing automatic verification of emissions, is helpful for this task. This relieves the operator by freeing him of the necessity to continuously check for the presence of the 'correct' signal.

Modern *classification receivers* allow the largely automated recognition of the modulation type and possibly the coding method used (Fig. II.29). In state-of-the-art systems they are no longer separate stand-alone units, but integrated receiver modules of the required functionality or software-controlled modules. Based on the contents of (self-learning) databases they enable future identification or recognition of emissions and may directly follow the current data of public allocations. By classifying the stream of symbols or the method used these examine the demodulated signal for characteristics that may serve to determine the method and possibly the code employed. Pattern-recognizing programs can relate the recordings to the time/frequency level or the waterfall display (Section II.4.2) based on the fact that the various transmission methods leave a characteristic footprint. For this purpose, the receive signals are evaluated automatically over a certain period with regard to the emission class and transmit frequency used (and possibly the direction of incidence). With A3E-modulated signals, for example, this is the carrier with

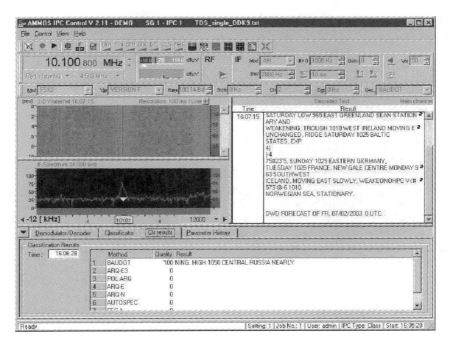

Figure II.28 A receiver path in the AMMOS radio intelligence system from Rohde&Schwarz classifies the parameters of the input signal (Baudot in this example) and then demodulates and decodes a Baudot radio weather signal for recording. (Company photograph of Rohde&Schwarz.)

its sidebands. Frequency-hopping emissions and burst emissions can also be recognized and differentiated automatically by their specific occupancy rate. The frequency-hopping method is characterized by the distribution of the spectral energy over a wide energy range containing several frequency pattern elements [24]. (When evaluating frequency-hopping emissions in connection with a direction finder, it may be of advantage to combine the frequencies received within a certain range of the angle of incidence.) Short emission times may indicate burst emissions.

Signal segments of a detected emission can be saved in a database [25]. The database can be used to combine several successfully acquired segments to form segment groups for later identification of the transmitting station or the transmitter classification. If at some later time a transmitter is detected to which an already known segment or segment group can be assigned, a relocation of that transmitter or a change in its transmit frequency can be assumed. Furthermore, it is possible to identify those transmitters that belong together in regard to their communications content and form transmitter associations or radio networks.

Recovering the contents of encoded emissions, called *production* in this area, is of prime importance for reconnaissance in intelligence or military applications. The signal is demodulated and the regained content is made audible or visible on the screen. Depending

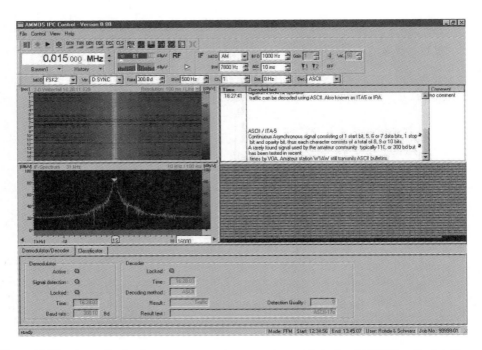

Figure II.29   According to its preprogramming a receiver path in the AMMOS radio intelligence system from Rohde&Schwarz automatically follows the change in signal coding from ASCII to ARQ-E3 in this example. First, the ASCII decoder recognizes that the signal can no longer be processed and then the receiver path changes decoders until the signal can be decoded again (here by the ARQ-E3 decoder). The darker area below the text window shows the symbol stream in real-time. (Company photograph of Rohde&Schwarz.)

on the type of emission, this is (clear) text or an image. The cutting edge technology is described in [26] and [27]. Deciphering and decoding is performed by modern PC-supported procedures in combination with specialist's expert knowledge.

II.4.4.1   Problems with Emissions Occurring at the Same Time or at Nearly the Same Frequencies

In practice, the signals of several emitters may be superimposed as illustrated in Figure II.16. This phenomenon is called co-channel interference. Separating and segmenting only on the basis of amplitude distribution is very time consuming and prone to errors, especially when using automated methods. Direction finding (Section II.5.1) can help to discriminate between individual emissions by determining the direction of incidence. Emissions of a certain basic type and coming from the same direction are very likely to originate from one and the same emitter. On the basis of bearing results it is possible to perform the segmentation of (LPI) signals occurring at the same time and in the same frequency range. However, there are limitations owing to,

- too much overlap of emissions in the time-frequency spectrum (when signals appear 'superimposed'),
- insufficient coverage of the time-frequency spectrum by the direction finder (insufficient time and frequency resolution), or
- limited accuracy of the direction finder (DF) [16].

Weighed measurement values can also be derived from high-resolution waterfall presentations, as used by national authorities performing radio reconnaissance according to the ITU (International Telecommunication Union). These values may include:

- Frequency measurements, also in regard to variations over time, in the case of fixed-frequency emissions and to the frequency offset.
- Field strength measurements (Section III.18).
- Measurements of the occupied bandwidth and of spurious emissions close to the carrier frequency via the air interface.
- Measurement and documentation of modulation parameters (modulation depth (A3E) or modulation index and frequency swing (F3E)) and the coverage area of broadcasting stations via the air interface.

Important requirements are specified in the current edition of the ITU recommendations in [28] to [34]. The receiver path becomes virtually an *analysis receiver* taking over the measuring functions. It supplies typical technical parameters like the centre frequency, bandwidth, modulation type, and other parameters that depend on the type of equipment, like the shift, symbol rate, number of channels, channel spacing and burst length. The functions are increasingly similar to those of modern spectrum analyzers, so that this type of receiver is increasingly used for the tasks described. This is particularly true for the VHF/UHF/SHF range. However, both of these units have their advantages and are particularly suitable for their own field of use (Table II.6). A comprehensive survey can be found in [35].

II.4.5 Receiver Versus Spectrum Analyzer

Seen in terms of their conceptual design, spectrum analyzers follow basically the same construction as multiple-conversion heterodyne receivers. However, they function more and more in the way of a second generation digital receiver (Fig. I.16). Units called real-time spectrum analyzers have recently become available for use with frequencies up to the medium SHF range. The system design is similar to that of FFT multi-channel receivers [36].

Contrary to receivers designed for radio reconnaissance and radio surveillance, spectrum analyzers have a different operating concept that is adapted to the task to be performed (Table II.6). This is also the case for so-called black-box units, receivers that have no

Table II.6   Differences between receivers and spectrum analyzers according to [35]

Receiver for reconnaissance/surveillance	Spectrum analyzer
Direct buttons for applications like monitoring of	*Direct buttons for applications like measuring of*
– various search run functions	– sweep functions (span, centre)
– emission classes/demodulators	– reference line
– receive bandwidths (Section III.6.1)	– resolution filter (RBW, VBW)
– AGC acting time (Section III.14)	– trigger
– AFC (Section III.15)	– measuring detectors/weighting (rms, quasi-peak, peak, sample, etc.)
– input attenuator	– calibrated input attenuator
– storing data to/retrieving data from memory	– display mode (clear/write, average, max./min. hold, etc.)
In addition controls for	*In addition front inputs for*
– squelch	– optional tracking generator
– RF gain	– external power measuring head
– MGC (Fig. I.5)	– external harmonic wave mixer
– cursor (marker)	
The hardware concept requires	
– RF preselection for large-signal behaviour (Section III.12) and minimum receiver stray radiation (Section III.17)	– a YIG prefilter for image band suppression (Section III.5.3) from SHF on
And offers primarily	
– demodulators for AM/CW/FM/PM/ISB/SSB	– demodulators for (AM)/FM
– AGC	– measuring and weighting filters
– AFC	– linear/logarithmic level display
– automatic input attenuator	– calibrated input attenuator
– search run functions	– standardized overlay measuring mask
– squelch	– various trigger options
– AF filter	– optional cursor functions for typical RF measurements (IP3, dBc, etc.)
– only one LAN interface	– interfaces to peripheral units
– (two separate signal processing paths)	– one signal path for all functions
– separate setting of span and receive bandwidth	– high measuring accuracy
– high tuning speed	– relatively low wobble speed
And allows the display/measurement of	
– spectrum	– spectrum
– IF spectrum	– RF and possibly baseband parameters of specified standards (GMS, UMTS, etc.)
– video spectrum	– adjacent channel power (ACPR)
– pulse width	– RF-typical measurements (EVM, F_{dB}, etc.)
– pulse spectrum	– constellation diagrams
– requirements according to ITU recommendations	– measuring masks
– time dimensions	– time range (zero span)
– waterfall diagrams	– out-of-band signals (harmonic waves)
– histograms	
– decoding and deciphering results*	
– possibly constellation diagrams*	
Total measuring uncertainty up to 3 GHz	
– <3 dB with units of latest generation	– <0.35 dB with units of latest generation

*Mostly in connection with external offline analysis.

front panel and are controlled via a PC. These are primarily intended for fast access to functions like spectrum display, selection of emission classes and their demodulation, choice of receive bandwidth, etc., since the signals to be processed are often available for a short time only. The overall concept of these units focuses on real-time capability with few compromises!

The front end of the receiver incorporates a preselector assembly and attenuating elements variable in steps, and/or low-noise preamplifiers which are switched manually or automatically for adaptation to the receive signal scenario (Section IV.2). This type of automatic optimization of the RF path, combined with AGC (Section III.14), is important since the unattended operation and automatic search mode allow no manual switching. With the spectrum analyzer, on the other hand, the input attenuator circuit (for the adjustment of RF levels) is switched manually in fine increments or is linked to the manual selection of the reference line.

Receivers are generally optimized for operation with antennas and receive signals via air interface. In practice, the signals are often quite distorted (multi-path propagation, fading, etc.) and require special treatment, for example, equalizers, for subsequent successful processing. The spectrum analyzer is designed for measuring tasks in laboratories or test rooms. Usually it operates with conducted signals and not with antenna signals. Its primary job is the measurement and qualitative evaluation of known signals. Its demodulating capabilities are optimized for signals that are rarely or only selectively disturbed (e.g., defined superposition of noise). The demodulation algorithms used in a spectrum analyzer for signal analyses must also regenerate the ideal reference signal for error calculations (e.g., when measuring the error vector magnitude (EVM)). Signal processing is therefore more extensive [35].

II.5   Direction Finding and Radio Localization

Different methods of radio direction finding use the incident wave front to obtain information about the direction of the origin for the radio signal to be localized.

Two basic parameters are of prime interest. One is the *azimuth*, which indicates the horizontal direction of incidence either relative to the observer's location or absolute as the point of compass. The other is the *elevation*, which allows the determination of the distance between the direction finder and the transmitter, especially in the short wave range.

Identifying the direction is done either with the help of the directional characteristic of the antenna or by reconstructing the incident wave front from sampling values of the field distribution in space (in relation to the phase and the amplitude).

II.5.1   Basic Principles of Radio Direction Finding

According to the range and field of receiver applications a number of different methods and systems have evolved. These are largely adapted to the individual intended use. In order to obtain a deeper understanding of the receivers required for these applications they will now be discussed in regard to their fundamental design. Following the overviews in [20] the survey will be limited to the configurations currently most often used. More details can be found in [17] and [37].

II.5.1.1 Direction Finding through Directional Characteristics of Antennas

This method may be the best-known technique for establishing the direction of incidence. Depending on whether the highest or lowest voltage received is evaluated, it is called the maximum bearing or minimum bearing, whereby the minimum of the radiation lobe of a directional antenna is usually 'more prominent' and therefore delivers better results. Frame antennas or loop antennas are well suited for long wave lengths up to the SW range, while log-periodic antennas are often used in the VHF range.

By using a rotor to continuously rotate the antenna and by displaying the received voltages on a screen in relation to the angle of rotation, a graphic evaluation is obtained (Fig. II.30). The angle of incidence can be read directly (rotational direction finder). A combination of several antennas of the same type arranged in different positions provides the known cross bearing (also called triangulation or intersectional bearing). The advantages are the relatively simple construction and the possibility to use the same antenna for direction

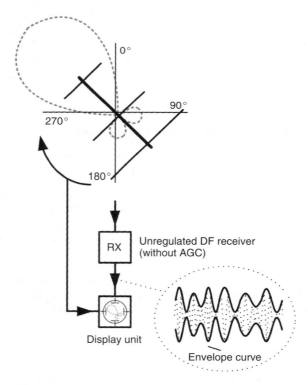

Figure II.30 Functional principle of the rotational direction finder: The unregulated signal is processed from the single channel DF receiver without AGC to indicate the direction of incidence.

finding and radio surveillance tasks. The system is also very sensitive, due to the type of antenna used. Especially in the microwave range this mechanical method of direction finding (DF) is regarded as a very good compromise.

II.5.1.2 Adcock Principle

In order to prevent the unwanted influence of polarization disturbances by sky waves (the incident polarization angle is different from the antenna polarization) the direction finder frame is replaced by antennas receiving mostly the vertical components of the incident electrical field. These usually consist of two symmetrical vertically oriented dipoles spaced apart at a distance smaller than the wavelength, since larger distances would cause side lobes in the antenna diagram, with the resulting ambiguities.

In the basic design of the Adcock direction finder the antenna system rotates. Because of the two receive minima that occur, the result remains ambiguous. Owing to the addition of the received voltage from an auxiliary antenna the direction of incidence can be clearly identified. In order to increase the rotational speed and to allow such systems to function with lower frequencies (larger antennas), two antenna groups offset by 90° have been used according to Adcock (Fig. II.31). These feed the receive voltages to a so-called goniometer. This consists of two orthogonally crossed coils with a search coil in the centre which rotates and provides the desired voltage minimum. The effect is the same as though the antenna were rotated mechanically. The voltage at the search coil is then picked up by the direction finding receiver. The advantages are less bearing errors and a much faster scan over the entire range of 360° since no antenna rotation is necessary.

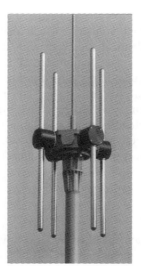

Figure II.31 Adcock antenna arrangement for the VHF frequency range. (Company photograph of PLATH.)

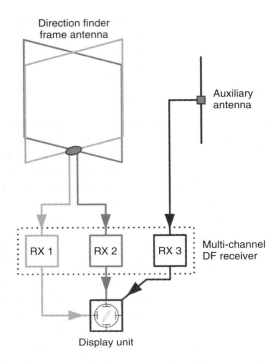

Direction finder
frame antenna

Auxiliary
antenna

RX 1   RX 2   RX 3   Multi-channel
DF receiver

Display unit

Figure II.32   Functional principle of the visual direction finder according to Sir Watson-Watt.

II.5.1.3   Watson-Watt Principle

When feeding the antenna output voltages (which may come from Adcock antennas or cross-frame antennas) via identical receivers with the same internal transit time to a cathode-ray tube instead of to the goniometer, a Lissajous pattern appears on the screen. This system is called a cathode-ray direction finder according to Sir Watson-Watt (Fig. II.32). In an ideal case, the pattern on the screen is a line at an angle that indicates the direction of the incident wave. If the above mentioned polarization distortions are present, the indicated line changes more and more to an ellipse. The quality of the bearing (clouded bearing) is directly available. The main advantage is the almost inertia-free operation of the system (short-time direction finder) and, in addition, the immediate indication.

II.5.1.4   Doppler Principle

If an antenna rotates with a certain speed around a centre, the receive signal is frequency modulated with the rotational frequency as a result of the Doppler effect.

> Rule: *If the receiving antenna rotates against the propagation direction of the incident wave front, the measured frequency is higher. If the antenna rotates with the propagation direction the measured frequency is lower!*

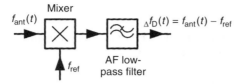

Figure II.33 Illustration of the mixing (Section V.4.1) of two signals – $_\Delta f_{\text{ant}}(t)$ with a frequency modulated by the rotation and the centre frequency f_{ref} as a reference. The latter is derived from the reference direction. The low-pass filter serves to remove other unwanted mixing products.

When mixing this frequency with a reference signal (adjusted receiving frequency, centre frequency) in the AF band, the remaining signal is the Doppler shift (Fig. II.33). The zero crossings of the signal can now be compared to a reference phase. This reference phase is obtained by specifying a reference direction for the rotating antenna. (With fixed direction finders this is usually the Northern orientation. With mobile direction finders it is often the longitudinal axis of the vehicle facing the engine.) Since both angles can have a maximum of 360° the phase difference corresponds to the angle of incidence!

Instead of continuously rotating the antenna, each of several fixed antennas arranged in a circular pattern is connected by an electronic switch to the input of a DF receiver. The antennas are than 'scanned' by means of a so-called commutator (Figs. II.34 and II.44). Owing to the increased rotational speed this results [9] in an easily evaluable Doppler shift:

$$_\Delta f_{\text{D}}(t) = \frac{\O \cdot \pi \cdot n \cdot \sin(2 \cdot \pi \cdot n \cdot t)}{\lambda} \tag{II.3}$$

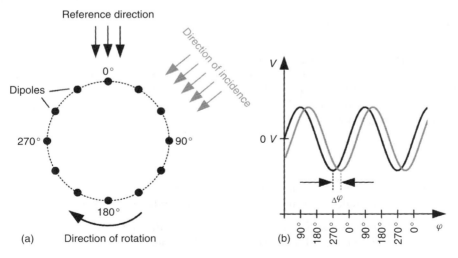

Figure II.34 Functional principle of a Doppler direction finder. (a) Receiving dipoles in circular arrangement. (b) Phase shift $_\Delta \varphi$ dependent on the angle of the incident wave front.

where

$_\Delta f_D(t)$ = Doppler shift at time (t), in Hz

$Ø$ = diameter of the circular antenna arrangement, in m

n = antenna rotational frequency, in Hz

t = time considered, in s

λ = operating wave length, in m

The maximum shift is the frequency swing

$$_\Delta f_{D\,max} = \frac{Ø \cdot \pi \cdot n}{\lambda} \qquad (II.4)$$

where

$_\Delta f_{D\,max}$ = maximum swing of Doppler shift, in Hz

$Ø$ = diameter of the circular antenna arrangement, in m

n = antenna rotational frequency, in Hz

λ = operating wave length, in m

The diameter of the circular antenna path can be of any size. But the distance between two individual antenna elements must be much smaller than half the operating wave length [17]. This allows the bearing basis to be so large that only small bearing errors occur even in the case of multiple incident waves. This is due to multipath propagation. In addition, a large bearing basis enables high DF sensitivity, which means high accuracy, in recognizing the direction of incidence with low receive field strengths (Section III.18). In some cases, the reference phase is continuously available from a separate fixed reference antenna feeding a second channel of the direction finding receiver.

The most important advantages are high immunity against multipath propagation and high DF sensitivity (which is often improved by using active antenna elements). With typical rotational speeds of 10 to 200 rotations per second, these designs are now considered to be rather slow.

II.5.1.5 Interferometer Principle

With an interferometer the angle of incidence is determined by the direct measurement of the phase difference between several antennas arranged in a spatial pattern. This is basically a delay time measurement. In other words, from the phase differences of three antennas in their known geometric relation to each other the angle of the incident wave front can be calculated by

$$\alpha = \arctan \frac{\varphi_2 - \varphi_1}{\varphi_3 - \varphi_1} \qquad (II.5)$$

where

α = azimuth angle of the incident wave, in degrees

$\varphi_1, \varphi_2, \varphi_3$ = phase angle at the respective measuring probe, in degrees

If the azimuth angle is known, the vertical angle of the incident wave front can be determined from the measured antenna phases and the operating wavelength.

$$\varepsilon = \arccos \frac{\sqrt{(\varphi_2 - \varphi_1)^2 + (\varphi_3 - \varphi_1)^2}}{\dfrac{2 \cdot \pi \cdot d}{\lambda}} \qquad (II.6)$$

where

ε = elevation angle of the incident wave, in degrees

$\varphi_1, \varphi_2, \varphi_3$ = phase angle at the respective measuring probe, in degrees

d = distance between antennas 2 and 1 or 3 an 1, in m

λ = operating wavelength, in m

Owing to the necessary mathematical operations, interferometer systems became established with the introduction of digital signal processing.

On the one hand, a clear determination of the direction is possible only if the distance between the antennas is smaller than half the wavelength. On the other hand, the antenna array (bearing basis) should be as large as possible to achieve high accuracy. To satisfy these contradictory requirements over a wide frequency range, the antennas are arranged as shown in Figure II.35. At any given time, only the three antennas providing the best

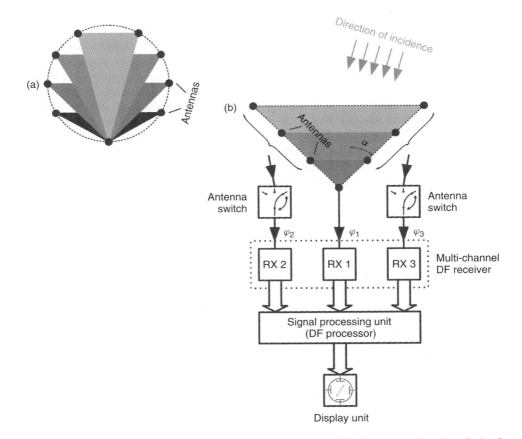

Figure II.35 Possible antenna arrangement of a three-element interferometer direction finder for use over a broad frequency range. They are placed (a) in a circular pattern or (b) in antenna rows. The antennas used in direction finding must be spaced at a distance of less than half the wavelength of the receiving frequency. Sometimes, the direction of incidence is determined without any doubt from a smaller antenna basis (Table II.5). Better resolutions can be achieved by measuring the phase differences with higher accuracy after changing to a larger bearing basis.

results are active. One of the channels of the direction finding receiver is permanently connected to the reference antenna used as a standard to identify the phase angles. Two more channels of the direction finding receiver are connected to the other antennas, depending on the signal of interest. The true angle of incidence is derived from the plus or minus sign of the phase difference.

One of the advantages is the fact that the azimuth and elevation can be determined directly. The directional characteristics and the polarization of the antenna elements have no influence on the result as long as they are the same for all elements [19], that is, the bearing angle is derived only from the phase relations between the antenna elements.

II.5.1.6 Problems with Direction Finding Aboard a Ship

The navy is faced with having to establish the direction of a signal from aboard a vessel. In radio reconnaissance missions this entails dealing predominantly with short-wave signals (Section II.4). Disturbances in the received wave field require methods that take into account the actual behaviour of DF antennas in the close vicinity to the ships superstructure. An analysis [38] shows that the ambiguity caused by the ship itself is the greatest contribution to the overall error. As a rule, the DF antenna system is arranged in the upper section of the ship's mast. A monopole or dipole is used for the vertically polarized auxiliary antenna that is essential (as described above), since a cross-frame antenna alone is unable to deliver non-ambiguous directional results. A specific problem with antennas installed at a certain height is the number of zeros in the elevation diagram that occur at small elevation angles in combination with wavelengths of approximately twice the mast length. This allows only unsatisfactory reception of horizontal waves. The large phase variations associated with this often induce 180° directional errors. Since shifting the antenna mechanically is helpful only within a narrow frequency range, a solution can be the installation of several antennas arranged at different heights, allowing antenna selection according to the operating frequency. The classical approach to HF direction finding on ships is the use of the Watson-Watt method. In an ideal case, the signals produce no phase shift and the bearing is derived directly from the sine or cosine azimuth dependency of the antenna signal voltages. In a distorted wave field, on the other hand, the antenna signal contains higher order modes which can lead to ambiguities. (The term 'mode' describes the form of propagation of an electromagnetic wave.) Furthermore, there are phase shifts affecting the direction finding results. The errors can be minimized by the method of evaluation. It appears to be reasonable to also use the higher modes and the azimuth-dependent phase shifts for obtaining the bearing. Since in a disturbed wave field the relationship between azimuth and amplitudes or phases of the antenna voltages depends on the actual design of the vessel, direction finding must be performed by comparison with sample data (correlation method). It has proved useful to use both methods alternatively: The Watson-Watt method is more efficient in frequency ranges with little field disturbances, since it takes the known antenna characteristics into account, while the correlation method can reduce the errors more efficiently in heavily disturbed frequency ranges since it uses the additional information contained in the distorted antenna characteristics [38].

II.5.1.7 Problematic Co-Channel Interferences

The classic direction finding methods are based on the assumption that only one dominating wave is seen in the frequency channel of interest. If this is not the case, bearing errors will occur because,

- several useful waves to be monitored are spectrally overlapping (like CDMA, see Fig. II.18),
- there are additional interfering waves of high amplitude (like electric interferences) in addition to the useful wave,
- multi-wave propagation (like reflections from buildings) exists.

This causes incorrect results. In the classic methods of direction finding [39] there are two known approaches to solve this problem:

(a) If the interfering wave section is of a lower magnitude than the useful section, the bearing error can be minimized by suitable dimensioning of the direction finder, especially by choosing a sufficiently large antenna array.
(b) If the interfering wave section is greater than or equal to the useful wave section, the bearing of non-correlated signals can be taken separately with high-resolution multi-channel direction finders based on FFT multi-channel receivers (Section II.4.2) by evaluating spectral differences.

Super-resolution DF methods offer a systematic solution of the problem by calculating both the number of waves involved and their angle of incidence with either the model-based maximum likelihood method or the principal component analysis (PCA) method. Such techniques are suitable for breaking down a wave field containing several signals of the same frequency. (If M is the number of antenna elements used by the direction finder and is equal to the number of DF channels, the method allows the separation of $M - 1$ waves.) The maximum number achievable depends on the angle of incidence and the signal-to-noise ratio (Section III.4.8) [39].

II.5.1.8 Problems with Signals in Fast Time Multiplexing

Similar difficulties to those of multi-wave incidents can occur in TDMA networks (Section II.4.1 and Fig. II.17). In fact the number of simultaneously incoming waves is not higher, but they follow one immediately after the other with the same frequency within the given frame and time slot (Fig. II.36). For determining the angle of incidence, this means that each time slot must be checked for the presence of signals (there are of course always unoccupied time slots in the frame) and that an angle of incidence must be assigned to the respective time slot. Furthermore, the signal (burst) within a time slot can be used for measurements only if it is not disturbed by other signals that may be superimposed due to different propagation times. This is discussed in [40] and a solution is suggested. To make a DF receiver suitable for signals in TDMA networks, the frame clock or the time slot frequency must be known. This information can be obtained by observing the

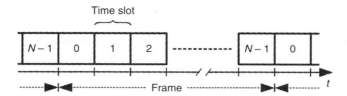

Figure II.36 Illustration of the frame clock and time slot frequency in TDMA networks. Several participants share the same frequency by making use of it only within the assigned time slot (Fig. II.17). Typical radio systems with TDMA access are cellular systems like the GSM (global system for mobile communications). Standard DECT (digital enhanced cordless telecommunication) uses a combination of TDMA and FDMA (Fig. II.15) [18].

emissions of the base transceiver station (BTS) within the cell of a cellular network, i.e. the downlink. Once the frame clock and time slot frequency are known, the angular values can be assigned to the network participants presently active. The frequencies can also be determined by observing the active participants (uplinks) in the cell.

For deducing the frame clock and time slot frequency the DF receiver is provided with another receive channel. Figure II.37 shows a DF receiver that obtains the clock information from BTS emissions and determines the angle of incidence by utilizing the interferometer principle described above. The receiver assigns an angle of incidence to every occupied time slot. This information is then forwarded together with the frequency, time and reliability of the angle information for further processing. If no reliable bearing is found in a time slot, this information is also forwarded for processing.

The bearing angles assigned are considered statistically significant if they are determined within one burst in the following way: sample values are taken simultaneously from all DF channels. Together these form a so-called snapshot. A bearing angle is calculated from each snapshot. Statistically independent snapshots provide statistically independent estimations of the bearing angle. A limiting condition for obtaining independent snapshots is that they are taken in certain time intervals:

$$t_i(B_{-6\,\text{dB}}) > \frac{1}{B_{-6\,\text{dB}}} \tag{II.7}$$

where

$t_i(B_{-6\,\text{dB}})$ = required time interval of snapshots for the used bandwidth $(B_{-6\,\text{dB}})$, in s

$B_{-6\,\text{dB}}$ = receive bandwidth ($-6\,\text{dB}$ bandwidth) of the receiver paths, in Hz

These are directly related to the bandwidth of the evaluated signal. The angles of incidence measured within a time slot are used to calculate the mean value and the variance. As long as the variance does not exceed a certain threshold, the mean value is the statistically significant bearing angle of the time slot and the variance is the degree of reliability of the direction determined [40].

Depending on the DF method the *direction finding receiver* uses one or several parallel receiving paths for amplification and selection of the signals received via the antennas used for determining the angle of incidence. (This are sometimes called direction finding

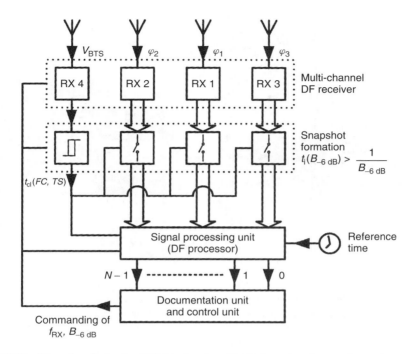

Figure II.37 Direction finder for TDMA signals with N time slots in one frame (as an example of the interferometer principle). The frame clock and time slot frequency ($t_{cl}(FC, TS)$) result from the signal V_{BTS} received from a base station. Processing the direction finding algorithm and the subsequent derivation of a reliable bearing are performed in the signal processing unit. The combined assignment of receiving frequency f_{RX} and receive bandwidth $B_{-6\,dB}$ (Section III.6.1) facilitates the process [40].

converter (DF converter).) The simple design of a *single-channel DF receiver* consists of a suitable receiver coupled to a direction-finding attachment. The latter contains the DF processor used for combining and processing the signals (filtering, bearing calculation, demodulation) and the display unit.

A multi-channel direction finding receiver usually feeds an LO injection signal (for tuning purposes) to all receiving paths to ensure accurate synchronization. In the individual channels the signals of the same frequency should be synchronized in amplitude and phase. Receivers of this type are used for multi-channel radio location, which can operate based on either the interferometer principle or the Watson-Watt principle (Fig. II.38). (According to [41] they require a phase synchronization of better than 1° and an amplitude synchronization of better than 0.1 dB.) In order to correct any synchronization differences between the channels, multi-channel DF receivers often use calibration methods that either manually or automatically detect and correct the channel synchronization errors by reference measurements in which, for example, all channels produce the same signal. With the increasing digitization of signal processing on the receiving side (Fig. I.20) these problems become less important. Digital filters with equal filter coefficients show exactly

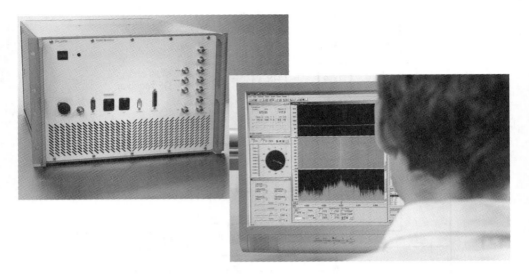

Figure II.38 The DFP 2400 compact VHF/UHF direction finding receiver (including DF processor and operating/display software) from PLATH. The device is intended for a frequency range from 20 MHz to 3 GHz (optionally to 5.8 GHz) and can operate in accordance with the Watson-Watt principle or the interferometer principle, depending on the type of antennas used. It also allows TDMA signal bearings. The analog receive bandwidth is fixed to 20 MHz. For different broadband signals (Section II.4.1) several surveillance widths of 20 MHz to 800 MHz can be selected by automatic tuning and shifting of the '20 MHz window' in combination with FFT. (Company photograph of PLATH.)

the same pattern in both magnitude and phase, not only in their passband but also at their filter slopes. If the A/D converter of a multi-channel receiver of this design operates in multiplexing mode, the receive channels will be free of synchronization errors from this interface onward. This means that channel synchronization errors will occur only in components upstream of the A/D converter like antennas, RF amplifiers, and mixers and are therefore independent of the filter bandwidth set in the digital filter. By introducing a correction factor in the digital filter it is possible to allow for the transfer function of the upstream components. This is not possible in analog filters [41].

As a general rule, DF receivers have to meet the same high demands regarding the performing characteristics (Part III) as receivers used in radio reconnaissance and radio surveillance. A sufficiently high near selectivity (Section III.6) is a particularly important design criterion.

II.5.2 Radio Reconnaissance and Radio Surveillance

In order to achieve optimum security and uninterrupted readiness for use in the most demanding situations (technically and operationally), there are very specific requirements for such systems. How broad the range would be for obtaining optimum results in direction finding can be seen from the list below [20]:

- high accuracy and/or high resolution;
- high sensitivity (Section III.4.5) with good large-signal behaviour (Section III.12) to allow the localization of weak signal sources with simultaneous strong interference carriers;
- many different receive bandwidths (Section III.6.1) for optimum adaptation to the bandwidth of the signals of interest as well as for maximizing the signal-to-noise ratio (Section III.4.8);
- immunity against multi-directional propagation, interferences, and polarization errors;
- shortest possible direction finding times to facilitate recognition of the bearing angle of brief signals like frequency-hopping signals (Section II.4.1) with a high degree of certainty;
- a frequency range as wide as possible; and
- sophisticated remote control for unmanned operation and automated processes which today are often inevitable.

As already discussed in Section II.4, direction finders are frequently used as functional elements in a system group of radio reconnaissance and radio surveillance equipment. While in the beginning the goal was to achieve useful results with few resources, monitors with cathode-ray tubes are now no longer in use for directional representations. Today a large part of the system can be remotely controlled from the LC display or PC screen. This also allows the display of an electronic map in the background with the indicated direction of incidence superimposed. The receive field strength (Section III.18) is indicated in whatever scale is required. From the panorama detected by the search receiver

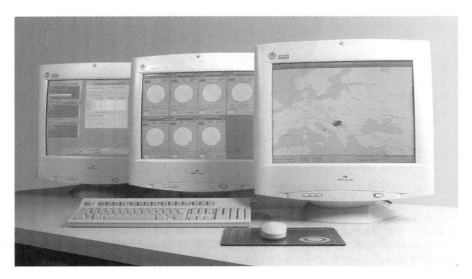

Figure II.39   Central control station for networks of several decentralized direction finders. The figure shows the OPAL system from PLATH as an example of an up-to-date workplace, showing (from left to right): The monitor for controlling the management tasks and for remotely controlling the field stations, the monitor for displaying the graphic results of the remote-controlled direction finders and the monitor for indicating the transmitter position. (Company photograph of PLATH.)

(Section II.4.2) a certain frequency can be selected with a movable cursor and one or several direction finders can be set to this frequency. The direction finders pick out one or several signal samples and forward the results (the direction determined) to a central station, where the individual results are combined to *localization results*, i.e. the radio location (Table II.5). These can be displayed as the transmitter location on an underlying map (Fig. II.39). Frequency tuning of the direction finder can be performed manually by an operator or automatically for each frequency detected by the search receiver [22]. In combination with other functional elements, such as those discussed in Sections II.4.2 to II.4.4, several reconnaissance missions can be processed in parallel.

Independent portable (hand-held) direction finders are used in the field. Such devices consist of a portable antenna, typically of log-periodic design or in the form of a loop antenna with switchable amplifier (making it an active antenna) in combination with a *portable monitoring receiver* or spectrum analyzer (Figs. II.40 and II.41). The strength of the received signal and thus the receive field strength is evaluated. Hand-held direction finders are used, for example, for detecting and localizing interference sources, micro-transmitters (usually referred to as 'bugs') or for tracing unauthorized emissions of vaguely known locations in urban surroundings (like localizing the floor in a building) (Fig. II.42). Another typical application is the localization of a distress alert signal source.

Figure II.40 The H500 RFHawk spectrum analyzer from Tektronix, together with a separate antenna, is specifically optimized for local signal tracking. The user interface allows superimposing local maps of the surroundings. By means of a GPS-based protocol [9] the device documents its own movements and provides a graphic display of the signal levels of the emission to be localized, measured at different locations. This allows detecting and avoiding faulty measurements taken at a certain location due to multiple signal propagation. The receiving frequency covers the range from 10 kHz up to 6.2 GHz. Maximum display width (span): 20 MHz; refresh rate of the spectral image: approximately 20 times per second; AM and FM demodulation. The dimensions are: $255 \times 330 \times 125$ mm. (Company photograph of Tektronix.)

Figure II.41   HE300 hand-held direction finding antenna together with the PR100 portable monitoring receiver from Rohde&Schwarz. Four different antenna modules can be plugged, so the entire frequency range between 9 kHz and 7.5 GHz can be covered. Maximum display width (span): 10 MHz; refresh rate of the spectral image: 20 times per second; AM, CW, FM, ISB, SSB, PULSE and IQ demodulation. The dimensions are: $192 \times 320 \times 62$ mm. (Company photograph of Rohde&Schwarz.)

Figure II.42   The 2261A Analyze-R^{TM} so-called Spectrum Monitor/Analyzer from Pendulum Instruments is primarily intended to detect local interferences in license-free and commonly used frequency bands (Table II.1). In combination with a GPS and external software it allows the documentation (logging) of receive levels and signal-to-noise ratios of signals received at various locations. With this method, the service coverage of a field station or one under construction can be tested. Receiving frequency range: 890 MHz to 940 MHz, 2.4 GHz to 2.5 GHz, 3.4 GHz to 4.2 GHz, 4.9 GHz to 6 GHz; display width (span): 100 MHz with 1 MHz resolution; refresh rate of spectral image: approximately 3 times per second; no demodulation capability. The dimensions are $89 \times 213 \times 333$ mm. (Company photograph of Pendulum Instruments.)

II.5.3 Aeronautical Navigation and Air Traffic Control

Air traffic control serves for the organizational and technical monitoring, planning and executing of flights [17]. The objective is to guide aircraft to their destination in a safe and timely manner. Long before the introduction of radar, air traffic control used direction finders very extensively. In situations of high air traffic density this helped to stagger the planes in an angular pattern. Today, direction finders are still of valuable assistance to flight controllers, since they complement the radar equipment. This applies especially to cases of aircraft flying outside (below) the detection range of the primary radar and of smaller aircraft not equipped with a transponder that produce only a very weak radar echo. Direction finders are important above all under visual flight conditions. If a pilot loses his bearings, flight control can radio him the directional details relative to magnetic north to follow in order to reach the landing strip. The aircraft itself requires no more than voice radio and compass, which are in any case mandatory.

Direction finding usually utilizes voice radio communication (Section II.3.1) between the aircraft and control tower. In addition to displaying the results on the direction finder monitor, the bearing is also displayed on the radar screen for flight identification (Fig. II.43).

Figure II.43 A synthetically generated radar image for the head of air traffic control also includes DF information. The demodulated AF signal from one of the A3E voice radio receivers shown in Fig. II.6 is also inserted. (Company photograph of Austro Control.)

For quite some time, direction finders based on the Doppler principle have been preferred for this task. It is advantageous to arrange the required antennas for VHF and UHF above one another instead of in a circular grouping (Fig. II.44). Even in complex and unfavourable scenarios they provide relatively accurate bearings.

In emergency aviation situations direction finders are of particular importance to air traffic control, since they can localize distress signals from emergency locator transmitters (ELT). These devices automatically emit a predetermined signal on reserved emergency frequencies of 121.5 MHz and 243 MHz. For energy saving reasons this is a pulsed signal.

For self-navigation and self-location of aircraft by means of non-directional beacons (NDB – described in Section II.3.1) automatic direction finders (ADF) are used. These are a part of so-called navigation receivers. Owing to the flat design required they are equipped with ferrite frame antennas arranged 90° offset to each other. An additional auxiliary antenna is necessary for the unambiguous determination of the angle of incidence. Integrating the device in an aircraft requires a certain level of sophistication in order to limit the directional errors due to aircraft structures to a minimum [17].

Here as well, simple GPS-based units [9] are used for self-navigation, especially in private and hobby aircraft.

Figure II.44 Antennas of a Doppler direction finder for air traffic control. Two systems are arranged above one another in order to obtain sufficient DF sensitivity in the frequency range monitored.

II.5.4   Marine Navigation and Maritime Traffic

Besides self-navigation, larger ships must have the ability to navigate to the port of destination by direction finding on the emergency frequency of 2,182 kHz [17]. The antennas used for this purpose are usually loop-cross frame antennas, either with a separate auxiliary antenna in the centre of the cross frame for unambiguously identifying the angle of incidence or with the two frames connected so that the received voltage is suitable for directional interpretation, making a separate rod antenna unnecessary. Here again, however, GPS-based devices [9] are often employed for self-navigation. Despite the use of navigation radio systems (Section II.3.2) for position indication and GPS-based means, direction finding is the only method that does not require extra auxiliary equipment, since there is always a large number of transmitters in operation at known locations and on known frequencies.

When taking the bearing of ships from on-shore stations, the position determined is usually quite accurate. The reference direction is more precise (since there are no on-board compass errors) and there are no bearing errors due to the ship superstructure in the near field of the DF antenna. Unlike aboard ships, a land-based DF station has more space for a wider antenna array. DF stations combined to form a location network can determine the position of a ship at the request of the crew. The ship sends the request to the control centre for execution. External radio location is also of importance in emergencies, especially if the ship is no longer capable of establishing its own position.

Furthermore, radio direction finding is used for marine traffic control and waterways near the coast, harbours and estuaries (Fig. II.45). With direction finders the marine traffic control centres have another valuable tool along with primary radar, which does not provide identification information. Apart from the indication of the DF monitor the identification

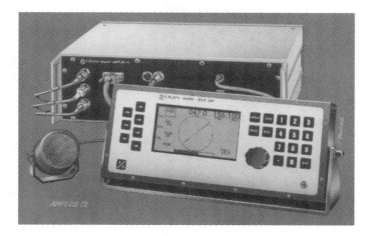

Figure II.45 AMPLUS 12 automatic three-channel direction finding receiver from PLATH (including DF processor and DF display unit), used predominantly in maritime and aeronautic traffic control with a frequency range of 108 MHz to 410 MHz. This type of equipment is employed in large numbers by the British Coast Guard. (Company photograph of PLATH.)

of the ship obtained by direction finding is also inserted in the radar display. This enables the differentiation of radar targets [17] during radio communication (Section II.3.2).

II.6   Terrestrial Radio Broadcast Reception

Various frequencies and different technical specifications are used for terrestrial radio broadcasting (Table II.7). Radio stations using the LW/MW broadcasting bands are spaced in a 9 kHz pattern, while those using SW broadcasting bands have a frequency spacing of 10 kHz and the classical VHF stations a spacing of 300 kHz. The typical *broadcast receivers* (or commonly radio receivers) are often supplemented by a simple integrated receiver for the MW frequency band with built-in ferrite rod antenna. An important criterion is very low harmonic distortion from demodulation (Section III.13.3), which guarantees a sound reproduction quality with a high signal-to-intereference ratio (Section III.4.8). The control elements are limited to a simple operating concept adapted to the typical broadcast listener. Contrary to the 50 Ω technology used presently in most RF applications, radio broadcasting still utilizes a 75 Ω receiver input impedance (Section III.3) on the receiver side if an external antenna connection is available. This is mostly for historic reasons in order to allow the adaptation of existing equipment to newer devices.

Simpler receivers designed primarily for the reception of several short-wave broadcasting bands are called *multiband receivers* or *world receivers*. They have built-in antennas, often of low efficiency, rather short rod antennas, or random wire antennas. Depending on the type of unit, these also cover other broadcasting bands. They are used, for example, by people travelling abroad and by emigrants to receive radio programs from their homeland or to listen to information or cultural news from far-away regions. Another typical application is to provide the local public in politically troubled areas with independent information. Users demanding higher standards also use *telephony receivers* (Section II.3.3) or *all-wave receivers* with separate receiving antennas (as detailed in Sections II.3.2 and II.3.4). Owing to their better receiver characteristics (Part III) and sophisticated demodulators they provide reception results of much higher stability and quality. Their superiority is also due to the receive bandwidth (Section III.6.1), which is matched to the respective received signal scenarios, provided that the receiver used offers a choice of different receive bandwidths. It may also be advantageous to demodulate an A3E-modulated radio broadcast in the J3E receiving path, since – if the interference affects one sideband only – demodulating the non-distorted sideband enables much clearer reception.

The term 'digital radio' describes the transmission of radio programs in digital broadcasting mode. International plans exist to eventually replace analog systems by digital technology in the coming years. With the orthogonal frequency division multiplexing (OFDM) based multi-carrier methods used, each carrier is modulated in amplitude and phase. The binary bit sequence of the individual carriers is finally modulated by quadrature phase shift keying (QPSK), differential quadrature phase shift keying (D-QPSK), or a higher-level quadrature amplitude modulation (QAM). Table II.7 indicates when which modulation method is used.

Digital audio broadcasting (DAB) [42] optimized for mobile reception will eventually replace all analog VHF FM broadcasting. The terms '*DAB receiver*' and '*DAB radio*' are

Table II.7 Terrestrial audio broadcasting frequency bands according to ITU RR [2]

Range	Band	Frequency (ITU RR region 1)	Frequency (ITU RR region 2)	Frequency (ITU RR region 3)	Modulation
LF	2 km–1 km	148.5 kHz–283.5 kHz			AM or OFDM with 4/16/64 QAM (DRM)*
MF	570 m–190 m	526.5 kHz–1606.5 kHz	525 kHz–1705 kHz*	526.5 kHz–1606.5 kHz	AM or OFDM with 4/16/64 QAM (DRM)*
MF	120 m	2,300 kHz–2,498 kHz	2,300 kHz–2,495 kHz	2,300 kHz–2,495 kHz	AM or OFDM with 4/16/64 QAM (DRM)
HF	90 m	3,200 kHz–3,400 kHz	3,200 kHz–3,400 kHz	3,200 kHz–3,400 kHz	AM or OFDM with 4/16/64 QAM (DRM)
HF	75 m	3,900 kHz–4,000 kHz*	3,900 kHz–4,000 kHz	3,900 kHz–4,000 kHz	AM or OFDM with 4/16/64 QAM (DRM)
HF	60 m	4,750 kHz–4,995 kHz	4,750 kHz–4,995 kHz	4,750 kHz–4,995 kHz	AM or OFDM with 4/16/64 QAM (DRM)
HF	60 m	5,005 kHz–5,060 kHz	5,005 kHz–5,060 kHz	5,005 kHz–5,060 kHz	AM or OFDM with 4/16/64 QAM (DRM)
HF	49 m	5,900 kHz–6,200 kHz	5,900 kHz–6,200 kHz	5,900 kHz–6,200 kHz	AM or OFDM with 4/16/64 QAM (DRM)
HF	41 m	7,200 kHz–7,450 kHz	7,300 kHz–7,450 kHz	7,200 kHz–7,450 kHz	AM or OFDM with 4/16/64 QAM (DRM)
HF	31 m	9,400 kHz–9,900 kHz	9,400 kHz–9,900 kHz	9,400 kHz–9,900 kHz	AM or OFDM with 4/16/64 QAM (DRM)
HF	25 m	11.6 MHz–12.1 MHz	11.6 MHz–12.1 MHz	11.6 MHz–12.1 MHz	AM or OFDM with 4/16/64 QAM (DRM)
HF	22 m	13.57 MHz–13.87 MHz	13.57 MHz–13.87 MHz	13.57 MHz–13.87 MHz	AM or OFDM with 4/16/64 QAM (DRM)
HF	19 m	15.1 MHz–15.8 MHz	15.1 MHz–15.8 MHz	15.1 MHz–15.8 MHz	AM or OFDM with 4/16/64 QAM (DRM)
HF	16 m	17.48 MHz–17.9 MHz	17.48 MHz–17.9 MHz	17.48 MHz–17.9 MHz	AM or OFDM with 4/16/64 QAM (DRM)
HF	15 m	18.9 MHz–19.02 MHz	18.9 MHz–19.02 MHz	18.9 MHz–19.02 MHz	AM or OFDM with 4/16/64 QAM (DRM)
HF	13 m	21.45 MHz–21.85 MHz	21.45 MHz–21.85 MHz	21.45 MHz–21.85 MHz	AM or OFDM with 4/16/64 QAM (DRM)
HF	11 m	25.67 MHz–26.1 MHz	25.67 MHz–26.1 MHz	25.67 MHz–26.1 MHz	AM or OFDM with 4/16/64 QAM (DRM)
(VHF	6.5 m	47 MHz–68 MHz*	54 MHz–72 MHz	47 MHz–50 MHz/ 54 MHz–68 MHz*)	
VHF	3 m (VHF-Band)	87.5 MHz–108 MHz	76 MHz–108 MHz	87 MHz–108 MHz	FM with preemphasis (Section III.4.9)*
VHF	1.5 m (Band III)	174 MHz–230 MHz	174 MHz–216 MHz	174 MHz–230 MHz	OFDM with D-QPSK (DAB)*
(UHF	64 cm–31 cm	470 MHz–960 MHz	470 MHz–608 MHz/ 614 MHz–890 MHz	470 MHz–960 MHz	OFDM with QPSK/16/64 QAM (DVB-T)*)
UHF	20 cm (L band)	1.452 GHz–1.492 GHz	1.452 GHz–1.492 GHz	1.452 GHz–1.492 GHz	OFDM with D-QPSK (DAB)

*Deviations/limitations according to country may apply.

Figure II.46   EVOKE-3 audio broadcast receiver from PURE Digital, which allows DAB reception in the 1.5 m and 20 cm bands as well as analog VHF FM reception. The unit can stop and rewind the DAB program. This receiver can record audio programs over 30 hours on a 2 GB SD memory card. From an internal 6-line graphic display the electronic program guide for one week can be displayed in order to select audio programs for recording. (Company photograph of PURE Digital.)

already in everyday use (Fig. II.46). The terrestrial digital video broadcasting (DVB-T) [43], developed for the transmission of digital TV programs, can also be used for digital audio broadcasting. Mixed channels are conceivable for the transmission of both TV and audio programs. Set-top boxes for DVB-T TV reception are commonly designed for the reception of additional DVB-T sound broadcasting. However, it remains to be seen whether or not DVB-T will actually be used for audio program transmissions.

Today, the importance of LW/MW/SW broadcasting bands for the typical audio program listener has declined somewhat. It is hoped that these will return with digital radio mondiale (DRM) [44], a digital standard for LW/MW/SW that enables the continued use (with some limitations) of the present 9/10 kHz spacing and the coexistence with current analog AM broadcasts. This provides the interesting opportunity of receiving the broadcast signal of a single audio transmitter over large areas and possibly over entire continents in stereo (with a *DRM receiver* (Figure II.47)).

Much to the contrary, sound broadcasting above 30 MHz requires booster stations to service a large area. *Relay receivers* or *re-broadcast receivers* serve to forward the signal emitted by the parent station from one relay station to the next. (Just like players of a ball game throw a ball, as a result of which some sources are called a 'ball receiver'.) The most outstanding characteristic of a ball receiver is the very low (linear and non-linear) modulation distortion and thus perfect left/right channel separation for stereo transmissions. Since they are typically placed in exposed locations, they often have a rather low sensitivity (Section III.4.8). All the more important are the high near selectivity (Section III.6), high blocking ratio (Section III.8), and clear frequency processing of the LO injection signal in order to minimize reciprocal mixing (Section III.7), since its several re-broadcast receivers can be installed within relatively small areas, so that the forwarded signals (programs) of different broadcasting stations reach high field strengths (Fig. II.48).

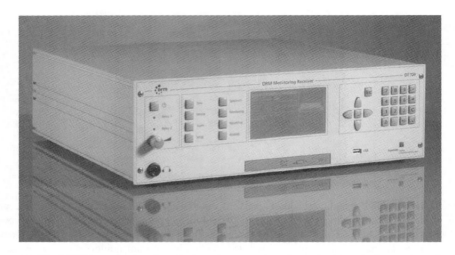

Figure II.47 DT700 professional audio broadcast receiver from the Fraunhofer Institute for Integrated Circuits, is used for low sound broadcasting bands ranging fom 2,020 m to 11 m. The receiver enables the demodulation of AM, DRM (completely according to [44]), and SSB modulation types. The unit operates as a digital receiver of generation 2.5 (Fig. I.19), that is, as a quasi all-digital receiver. From an HF input level of −110 dBm the DRM signal is demodulated or decoded. Owing to its performance parameters this receiver is suitable for use as a control receiver to monitor transmit signals at DRM transmitter stations. (Company photograph of Fraunhofer Institute for Integrated Circuits.)

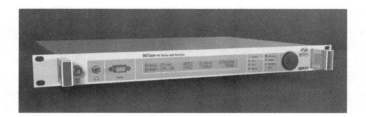

Figure II.48 B03 modern re-broadcast receiver from 2wcom, for example, for forwarding the audio program received with running an unmanned FM relay. (Company photograph of 2wcom.)

Broadcast stations use *control receivers* (measuring receivers) to continuously monitor the compliance of the emitted signal quality with the standards.

II.7   Time Signal Reception

The availability of an accurate time signal is of fundamental significance for many applications in daily life. In various countries like Germany, Japan, Russia, and the USA there are national institutes that provide exact time signals, which are picked up by suitable *time signal receivers*. Time signals can be used in equipment like radio clocks or time-based measuring devices after the exact time has been extracted from the signal.

These time signals are mainly transmitted by the LW band. As a long-wave signal, the time markers that are encoded by amplitude keying travel long distances. Small ferrite antennas are sufficient for their reception. Sometimes two ferrite antennas offset by 90° and connected via a transit time element are used for all-round reception.

Time signals are provided by time signal transmitters emitting a signal sequence according to a preset protocol. The various national time signal transmitters differ in both the selected transmit frequency and the structure of the protocol [45]. A good example is the LW transmitter DCF77 of the PTB, the federal German physical technical institute (Physikalisch-Technische Bundenanstalt), which receives time markers from several atomic clocks. The time signal is transmitted continuously on 77.5 kHz with 50 kW power. Other time signal transmitters include the BPC (China, on 68.5 kHz), HBG (Switzerland, on 75 kHz), JJY (Japan, on 40 kHz and 60 kHz), MSF (Great Britain, on 60 kHz), and WWVB (USA, on 60 kHz). The long-wave frequencies used are listed in Table II.8. The time information consists basically of the time signal within a time frame of exactly one minute length. This time frame includes values for the minute, the hour, the day, the week, the month, the year, etc. in the form of binary encoded decimal codes (BCD codes) emitted with pulse width modulation of 1 Hz per bit. Either the rising or the falling slope of the first pulse in the time frame is synchronized exactly to zero seconds. The time signal receiver of a typical radio clock is designed so that the time setting is realized by detecting the time information of one or several time frames from the moment the zero second signal is received for the first time.

Figure II.49 shows the coding diagram for the encoded time information following the protocol time signal transmitter DCF77. The coding formula comprises 59 bits, where 1 bit corresponds to one second of the frame. This means that in the course of one minute a time signal telegram containing binary encoded time and date information can be transmitted.

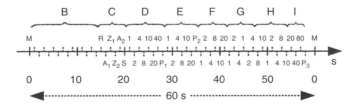

Figure II.49 DCF77 time signal telegram. The first 15 bits (sector B) contain a general code for operating information. The following 5 bits (sector C) contain general information: R is the antenna bit, A_1 is a bit signaling the changeover from central European time (CET) to central European summer time (CEST) and vice versa, Z_1 and Z_2 are zone time bits, A_2 is a bit for signaling a leap second, and S is the start bit for the encoded time information. The time and date in the BCD code are contained in the 21st bit to the 59th bit, with this data always applying to the following minute. The bits in sector D contain information on the minute, in sector E on the hour, in Sector F on the calendar day, in sector G on the weekday, in sector H on the month, and in sector I on the calendar year. This means that the information is encoded bit for bit. At the end of sectors D, E and I are the so-called parity bits P_1, P_2, P_3. The 60th bit is not occupied and serves to indicate the next frame. M represents the minute marker and thus the start of the time signal [45].

Table II.8 Terrestrial standard/time signal frequencies according to ITU RR [2]

Range	Band	Frequency (ITU RR region 1)	Frequency (ITU RR region 2)	Frequency (ITU RR region 3)
VLF	15 km	20 kHz (19.95 kHz–20.05 kHz)	20 kHz (19.95 kHz–20.05 kHz)	20 kHz (19.95 kHz–20.05 kHz)
VLF	12 km	25 kHz*		
LF	6 km	50 kHz*		
LF	4 km	72 kHz–84 kHz		
LF	3.5 km	86 kHz–90 kHz		
MF	120 m	2,500 kHz (2,498 kHz–2,502 kHz)	2,500 kHz (2,495 kHz–2,505 kHz)	2,500 kHz (2,495 kHz–2,505 kHz)
HF	75 m			4,000 kHz (3,995 kHz–4,005 kHz)
HF	60 m	5,000 kHz (4,995 kHz–5,005 kHz)	5,000 kHz (4,995 kHz–5,005 kHz)	5,000 kHz (4,995 kHz–5,005 kHz)
HF	37.5 m			8,000 kHz (7,995 kHz–8,005 kHz)
HF	30 m	10 MHz (9,995 kHz–10.005 MHz)	10 MHz (9,995 kHz–10.005 MHz)	10 MHz (9,995 kHz–10.005 MHz)
HF	20 m	15 MHz (14.99 MHz–15.01 MHz)	15 MHz (14.99 MHz–15.01 MHz)	15 MHz (14.99 MHz–15.01 MHz)
HF	19 m			16 MHz (15.99 MHz–16.005 MHz)
HF	15 m	20 MHz (19.99 MHz–20.01 MHz)	20 MHz (19.99 MHz–20.01 MHz)	20 MHz (19.99 MHz–20.01 MHz)
HF	12 m	25 MHz (24.99 MHz–25.01 MHz)	25 MHz (24.99 MHz–25.01 MHz)	25 MHz (24.99 MHz–25.01 MHz)

*Deviations/limitations according to country may apply.

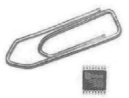

Figure II.50 MAS6179 integrated circuit for use in complete LW time signal receivers (here with sizes compared). This component offers the possibility to switch via a special input pin to three different transmitters. Additional external components required for its operation are only a ferrite antenna, three piezo-electric crystals for three different time signal frequencies, and a few capacitors. (Company photograph of MICRO ANALOG SYSTEMS.)

The emitted time signal is amplitude shift keyed with every second. The modulation consists of a decrease or an increase of the carrier signal at the beginning of each second. With the DCF77 transmitter the carrier amplitude is decreased at the beginning of each second for a duration of 0.1 seconds or 0.2 seconds to about 25% of the amplitude, with the exception of the fifty-ninth second of every minute. These decreases of varying length define the second markers or data bits. The change in the duration of the second markers serves for the binary coding of the time and date, whereby the second markers of 0.1 seconds duration represent binary '0' and those of 0.2 seconds duration represent binary '1'. Omitting the sixtieth second marker signals the next minute marker. In combination with the respective second, the time information can be derived by evaluating the transmitted time signal [45].

Since obtaining the signal requires the reception of only one fixed frequency, time signal receivers need no frequency tuning and are often of the TRF receiver type (Fig. I.2). Designed for one frequency only, these receivers produce useful reception results with a simple layout. Even a single passive circuit is often sufficient to achieve adequate selectivity in the LW range. Dedicated integrated circuits for time signal reception are commercially available (Fig. II.50).

II.8   Modern Radio Frequency Usage and Frequency Economy

With the continuously increasing demands on wireless communication there is a growing pressure on the availability of radio frequency resources. Yet [46] states that in all regions less than 20% of the frequency spectrum assigned by the authorities is actually used.

II.8.1   Trunked Radio Networks

Trunked radio networks, which were defined and realized by Telefunken, for example, for airport applications by the 1980s, represent simpler systems for the optimal usage of frequency economy. Prior to the actual voice radio communication each participant sends a brief data telegram to a central station. A free channel is then assigned for automatic selection by the participant's terminal equipment. In this way, the entire airport communication using dozens of hand-held radio units and mobile transceivers can be handled on less than 20 channels. This of course takes all emergency and priority calls into consideration.

Trunked radio systems [18] are also used for private and public company and security radio communications (called professional mobile radio (PMR)). Unlike conventional analog fixed-channel systems (with a certain radio channel assigned permanently to each of the services or participants), in trunked radio systems the frequencies are assigned *dynamically* to the respective services or participants. This allows using the advantages of 'trunking' and at the same time increasing spectrum efficiency.

II.8.2   Cognitive Radio

Cognitive radio (CR) is based on the principle of software-defined radio with its advantages as outlined in Section I.2.4, but is even more flexible. The term *'cognitive radio'* [47] describes an autonomous system which, in addition to its SDR functionality, monitors the current radio environment in order to identify and evaluate changes in regard to the frequency spectrum and its utilization and to respond to these if necessary. Receiving the signals and adapting to them the necessary processing can be done automatically with modulation type recognition. It is also possible to synchronize only to certain signals in the frequency spectrum received. More recently, the name cognitive radio also refers to entirely autonomously operating units (radio systems) featuring the respective functionality. The use of CR technology has been suggested in order to increase the efficiency of frequency spectrum utilization by introducing a dynamic management method for several frequency ranges (dynamic spectrum resource management). Here we distinguish between:

(a) Full cognitive radio – also known as 'Mitola radio': All parameters which can vary, for example, frequency and modulation, and which can be monitored, can be used for dynamic management.
(b) Spectrum-sensing cognitive radio: Here, the only variable parameter is the transmit frequency.

Such intelligent resource management methods in fact enable the user to occupy segments of already licensed frequency ranges not fully exploited by the initial licensee in all locations and at all times. The use of dynamic spectrum resource management can therefore allow the CR system to use unoccupied frequency segments, while taking the rights of the original licensee into account. According to [47] this requires that the CR system:

- knows its position (e.g., by means of global positioning system (GPS)),
- can evaluate the interference,
- observes the communication etiquette,
- is fair with regard to other users, and
- can inform the original licensee about its actions.

In addition, [48] points out that, in order to occupy unused spectral resources, the CR system must make use of a spectrum sampling technique for identifying the status of the present utilization of the spectrum quickly and accurately within a wide frequency range, also considering the different communication standards. So far, the spectrum sampling method can be divided into two groups:

(a) Energy detection – requires the careful selection of one or several threshold values and is often susceptible to noise or digitally modulated noise-like broadband signals.
(b) Feature detection – requires long preparation times compared with energy detection and requires extensive digital hardware resources, which (today still) results in high power consumption.

References

[1] Martin Sauter: Grundkurs Mobile Kommunikationssysteme – Von UMTS und HSDPA, GSM und GPRS zu Wireless LAN und Bluetooth Piconetzen (Basic Training Course on Mobile Telecommunications Systems – from UMTS and HSDPA, GSM and GPRS to Wireless LAN and Bluetooth Pico-Networks); 3rd edition; Vieweg & Sohn Verlag 2008; ISBN 978-3-8348-0397-9

[2] International Telecommunication Union (ITU), publisher: Radio Regulations; 2008 edition, Article 5 of The international Table of Frequency Allocations

[3] Thomas Rühle: Entwurfsmethodik für Funkempfänger – Architekturauswahl und Blockspezifikation unter schwerpunktmäßiger Betrachtung des Direct-Conversion- und des Superheterodynprinzipes (Methodology of Designing Radio Receivers – Architecture Selection and Block Specification with the Main Focus on Direct Conversion and Superheterodyne Designs); dissertation at the TU Dresden 2001

[4] Roland Bicker, Heinz Hagedorn, Severin Fischer, Christoph Saller: Neue Empfänger und zusätzliche Funktionen der Funk-Rundsteuerung (New Receivers and Added Functionality of Radio Ripple Control); ew 12/2004 – Vol. 103/23, pp. 34–37; ISSN 1619–5795

[5] Roland Bicker, Martin Eibl, Bernhard Sbick, Heinrich Wienold: Funk-Rundsteuersystem und Verfahren zum Betreiben eines derartigen Systems (Radio Ripple Control Systems and Modes of Operating such Systems); EFR Europäische Funk-Rundsteuerung 2002; patent specification DE10214146C1

[6] N. N. : Funk-Rundsteuerempfänger (Radio Ripple Receivers); LIC-Langmatz 1996; patent specification EP0726634B1

[7] European Radio Communications Committee (ERC)/European Conference of Postal and Telecommunications Administrations (CEPT), publisher: The European Table of Frequency Allocations and Utilisations covering the Frequency Range 9 kHz to 275 GHz; 1/2002

[8] Jochen Hinkelbein, Susanne Berger: Prüfungsvorbereitung für die Privatpilotenlizenz – Vol. 2 Beschränkt gültiges Sprechfunkzeugnis (Preparing for the Test to Obtain the Private Pilot Licence – Vol. 2 Limited Voice Radio Licence); 1st edition; AeroMed-Verlag 2007; ISBN 978-3-00-021004-4

[9] Werner Mansfeld: Funkortungs- und Funknavigationsanlagen (Radio Location and Radio Navigation Equipment); 1st edition; Hüthig Verlag 1994; ISBN 3-7785-2202-7

[10] Peter Iselt: System zum gemeinsamen Betreiben von auf verschiedene Wellenformen einstellbare digital arbeitende Funkgeräte (System for the Combined Operation of Digital Radio Equipment Adjustable to Different Waveforms); Rohde&Schwarz 1999; patent specification EP1201039B1

[11] Ulrich Graf: Lineares Frontend für den Eigenbau-RX – Teil 1 und Teil 2 (Linear Frontend for the Self-Made Receiver – Part 1 and Part 2); Funk Telegramm 9/1994, pp. 12–19, FunkTelegramm 10/1994, pp. 12–16

[12] Bernhard Walke: Mobilfunknetze und ihre Protokolle 1 – Grundlagen, GSM, UMTS und andere zellulare Mobilfunknetze (Mobile Radio Networks and Relevant Protocols – Basics, GSM, UMTS and Other Cellular Radio Networks); 3rd edition; B. G. Teubner Verlag 2001; ISBN 3-519-26430-7

[13] Bernhard Walke: Mobilfunknetze und ihre Protokolle 2 – Bündelfunk, schnurlose Telefonsysteme, W-ATM, HIPERLAN, Satellitenfunk, UPT (Mobile Radio Networks and Relevant Protocols – Trunked Radio Networks, Wireless Telephone Systems, W-ATM, HIPERLAN, Satellite Radio, UPT); 3rd edition; B. G. Teubner Verlag 2001; ISBN 3-519-26431-5

[14] Herbert Knirsch: Funkerfassungseinrichtung (Radio Monitoring Equipment); Rohde&Schwarz 1981; patent specification DE3106037C2

[15] Werner Kredel, Norbert Scheibel: Funkaufklärungssystem (Radio Intelligence Sytem); Daimler-Benz Aerospace 1994; patent specification EP0706666B1

[16] Franz Demmel, Ulrich Unselt: Erfassung und Peilung moderner Funkkommunikationssignale – Digitale Breitband-Suchpeiler R&S DDF 0xA (Detection and Direction Finding of Modern Radio Communications Signals – Digital Broadband Direction Finder R&S DDF 0xA); Rohde&Schwarz MIL NEWS 8/2004, pp. 18–23

[17] Rudolf Grabau, Klaus Pfaff, publishers: Funkpeiltechnik – peilen, orten, navigieren, leiten, verfolgen (Radio Direction Finding – Bearing-Taking, Locating, Navigating, Leading, Tracking); 1st edition; franckh Verlag 1989; ISBN 3-440-05991-X

[18] Michael Gabis, Ralf Rudersdorfer: Aktuelle digitale Funkstandards im transparenten Vergleich zum analogen FM-Sprechfunk – Teil 1 bis Teil 2 (Current Digital Radio Standards Transparently Compared with Analog FM Voice Radio – Part 1 and Part 2); UKWberichte 4/2007, pp. 195–208, UKWberichte 2/2008, pp. 107–119; ISSN 0177-7513

[19] Klaus Pfaff, Franz Wolf: Peil- und Ortungsanlage für Kurzzeitsendungen und zugehöriges Verfahren (Direction Finding and Localizing System for Short-Time Transmissions and Related Procedures); C. Plath Nautisch-Elektronische Technik 1993 (Nautical-Electronic Engineering 1993); patent application DE4317242A1

[20] Ralf Rudersdorfer: Zur Technik aktueller Funkortungs- und Funküberwachungsverfahren (On Current Technology of Radio Location and Radio Surveillance Equipment); manuscripts of speeches from the VHF Convention, Weinheim 2007, pp. 14.1–14.9

[21] Horst Stahl: Vielkanalpeiler (Multi-Channel Direction Finder); C. Plath Nautisch-Elektronische Technik 2000 (Nautical-Electronic Engineering 2000); patent application DE10016483A1

[22] Dieter Bienk, Gerhard Bodemann, Horst Ostertag, Helmut Schöffel, Jürgen George: Funkaufklärungsanordnung (Radio Intelligence Equipment Arrangement); Telefunken Systemtechnik 1988; patent specification DE3839610C2

[23] Paul Renardy: Vielkanalempfänger (Multi-Channel Receiver); Rohde&Schwarz 2004; patent application DE102004055041A1

[24] Gerhard Rößler, Horst Kriszio, Günter Wicker: Verfahren zur Feststellung, Erfassung und Unterscheidung von innerhalb eines bestimmten Frequenzbereichs unabhängig voneinander auftretenden Funksendungen (Method of Detecting, Recording and Discriminating Unrelated Radio Transmissions Occurring within a Certain Frequency Range); Battelle Institute 1982; patent specification DE3220073C1

[25] Marc Aguilar, Wolfgang Dolling, Peter Kropf, Franz Wolf: Verfahren zum Orten von insbesondere frequenzagilen Sendern (Method for Localizing Particularly Frequency-Agile Transmitters); C. Plath Nautisch-Elektronische Technik 1995 (Nautical-Electronic Engineering 1995); patent specification EP0780699B1

[26] MEDAV, publisher: Datenblatt OC-6040 Analyse-Arbeitsplatz für Übertragungsverfahren – frei konfigurierbar über 200 Verfahren (Datasheet on Workstation OC-6040 for Analyzing Transmission Methods – Freely Configurable for more than 200 Methods); Rev. 0/2008

[27] Thomas Krenz: Neues Identifizierungsmodul mit mehr als 120 Dekodierverfahren – Spectrum Monitoring Software R&S ARGUS (New Identification Module Including more than 120 Decoding Methods – Spectrum Monitoring Software R&S ARGUS); Neues von Rohde&Schwarz (Rohde&Schwarz News) III/2007, pp. 66–69; ISSN 0548-3093

[28] International Telecommunication Union (ITU), publisher: Automatic Monitoring of Occupancy of the Radio-Frequency Spectrum; ITU Recommendation SM.182 2/2007

[29] International Telecommunication Union (ITU), publisher: Spectra and Bandwidth of Emissions; ITU Recommendation SM.328 5/2006

[30] International Telecommunication Union (ITU), publisher: Accuracy of Frequency Measurements at Stations for International Monitoring; ITU Recommendation SM.377 2/2007

[31] International Telecommunication Union (ITU), publisher: Field-Strength Measurements at Monitoring Stations; ITU Recommendation SM.378 2/2007

[32] International Telecommunication Union (ITU), publisher: Bandwidth Measurement at Monitoring Stations; ITU Recommendation SM.443 2/2007

[33] International Telecommunication Union (ITU), publisher: Automatic Identification of Radio Stations; ITU Recommendation SM.1052 7/1994

[34] International Telecommunication Union (ITU), publisher: Technical Identification of Digital Signals; ITU Recommendation SM.1600 11/2002

[35] Christian Gottlob: Spektrumanalysator versus Kommunikationsempfänger (Spectrum Analyzer versus Communications Receivers); manuscript of speech for Rohde&Schwarz seminar series HF-Messtechnik & Digitale Kommunikation, Vienna 2007, pp. 1–42

[36] Tektronix, publisher: RF Signal Monitoring and Spectrum Management Using the Tektronix RSA3000B Series Real-Time Spectrum Analyzer; Tektronix Application Note 37W-21772-0 2/2008

[37] Friedrich K. Jondral: Einführung in die Grundlagen verschiedener Peilverfahren – Teil 1 bis Teil 2 (Introduction into the Basic Principles of Various Methods of Direction Finding – Part 1 and Part 2); ntzArchiv 9/1987, pp. 29–34, ntzArchiv 9/1987, pp. 67–72; ISSN 0170-172X

[38] Franz Demmel: HF-Peilung auf Schiffen (RF Direction Finding Aboard Ships); Rohde&Schwarz MIL NEWS 4/2000, pp. 7–10

[39] Philipp Strobel: Hochauflösendes Peilverfahren identifiziert Gleichkanalsignale – Digitale Überwachungspeiler R&S DDF0xA/E (High-Resolution Direction Finding Identifies Co-Channel Signals – Digital Surveillance Direction Finder R&S DDF0xA/E); Neues von Rohde&Schwarz (Rohde&Schwarz News) III/2007, pp. 72–73; ISSN 0548–3093

[40] Friedrich Jondral, Hinrich Mewes: Peilempfänger für den Einsatz in TDMA-Netzen (Direction Finding Receivers for Use in TDMA Networks); C. Plath Nautisch-Elektronische Technik 1997 (Nautical-Electronic Engineering 1997); patent application DE19701683A1

[41] Horst Stahl, Peter Fast: Verfahren zur Korrektur von Gleichlauffehlern von mindestens zwei Signalkanälen mit Digitalfiltern bei Funkortungsempfängern und Vorrichtung zur Durchführung des Verfahrens (Method for Correcting Synchronization Errors of at least Two Signal Channels Using Digital Filters in Radio-Locating Receivers and the Device for Performing the Method); C. Plath Nautisch-Elektronische Technik 1984; (Nautical-Electronic Engineering 1984); patent specification DE3432145C2

[42] European Telecommunications Standards Institute (ETSI), publisher: Radio Broadcasting Systems – Digital Audio Broadcasting (DAB) to Mobile, Portable and Fixed Receivers; ETSI Standard EN300401 6/2006

[43] European Telecommunications Standards Institute (ETSI), publisher: Digital Video Broadcasting (DVB) – Framing Structure, Channel Coding and Modulation for Digital Terrestrial Television; ETSI Standard EN300744 11/2004

[44] European Telecommunications Standards Institute (ETSI), publisher: Digital Radio Mondiale (DRM) – System Specification; ETSI Standard ES201980 2/2008

[45] Roland Polonio, Hans-Joachim Sailer, Christian Polonio: Programmierbarer Zeitzeichenempfänger, Verfahren zum Programmieren eines Zeitzeichenempfängers und Programmiergerät für Zeitzeichenempfänger (Programmable Time-Signal Receiver, Method of Programming a Time-Signal Receiver and Programming Device for a Time-Signal Receiver); ATMEL Germany, C-MAX Europe 2006; patent specification DE102006060925B3

[46] Michael Gabis: Funkkommunikation zur Einsatzunterstützung – Machbarkeitsstudie zum Einsatz neuer Funktechnologien bei Feuerwehreinsätzen (Radio Communications for Operational Support – Feasibility Study for Applying New Radio Technologies in Fire Fighting Operations); diploma thesis at the Johannes Kepler University Linz 2007

[47] Friedrich K. Jondral: Cognitive Radio – Environment Sensitive Mobile Terminals; manuscript of speech from the Heidelberger Innovationsforum, Heidelberg 2005, pp. 1–15

[48] Tajoong Song, Jongmin Park, Youngsik Hur, Kyutae Lim, Chang-Ho Lee, Jeongsuk Lee, Kihong Kim, Seongsoo Lee, Haksun Kim, Joy Laskar: Systeme, Verfahren und Vorrichtungen für eine Technik zur Erzeugung einer langen Verzögerung für das Abtasten des Spektrums bei CR (Cognitive Radio) (Systems, Methods and Devices for a Technique of Prolonged Delaying the Spectrum Scan in CR (Cognitive Radio)); Samsung Electro – Mechanics 2007; patent application DE102007035448A1

Further Reading

Erich H. Franke: Technologien zum Empfang breitbandiger Signalquellen auf Kurzwelle (Technologies for Receiving Wideband Signal Sources on Short Wave); manuscripts of speeches from the VHF Convention, Weinheim 1996, pp. 3.1–3.9

Jürgen Modlich: Klarer Durchblick im Gedränge der Signale – Funkaufklärungssystem R&S AMMOS (Complete Transparency despite the Density of Signals); Rohde&Schwarz MIL NEWS 7/2003, pp. 24–28

Wolfgang Schaller: Verwendung der schnellen Fouriertransformation in digitalen Filtern (Utilizing Fast Fourier Transformation in Digital Filters); NTZ 12/1974, pp. 425–431; ISSN 0948-728X

James Tsui: Special Design Topics in Digital Wideband Receivers; 1st edition; Artech House 2010; ISBN 978-1-60807-029-9

III

Receiver Characteristics and their Measurement

III.1 Objectives and Benefits

Antennas provide very different signal scenarios for receivers. Besides the strong and sometimes very weak useful signals, many other signals are fed to the receiver input. One of the main tasks for the radio receiver is to select the signal of interest and demodulate it in optimum quality to retrieve the information content. This process should be affected as little as possible by the other signals present, the interfering signals. Technical parameters are used to describe how a receiver performs in different situations. These so-called *receiver characteristics, receiver properties* or *receiver parameters* define the equipment efficiency and allow comprehensive objective comparisons. When comparing data from different sources, it is important that these parameters have been determined under the same conditions! Only this gives a meaningful and transparent comparison. While some specifications like receiver sensitivity (Section III.4) can be converted if established with different receive bandwidths (Section III.6), the comparison of other parameters is utterly useless if these are not measured by identical methods. By dividing the following text into two parts, the first of which describes the basic meaning of a receiver parameter and the other details of the measuring procedure required, the book will help to develop a feeling for the true significance of the specified characteristics.

Manufacturers are committed to maintaining the values guaranteed in their published specifications and data sheets. Typical values, on the other hand, only provide information about the average percentage with which certain data meet the specified value (e.g., in more than 98% of all units delivered). Such typical values of course appear better than the guaranteed values, since these must not necessarily be reached in the actual device. Manufacturers often specify only a limited number of characteristics. To obtain more information on the performance of a unit, it is necessary to conduct one's own test series or to read relevant test reports in professional magazines.

Radio Receiver Technology: Principles, Architectures and Applications, First Edition. Ralf Rudersdorfer.
© 2014 Ralf Rudersdorfer. Published 2014 by John Wiley & Sons, Ltd.

Section III.12 describes how some receiver characteristics interact during signal reception under real conditions and why they should not be looked upon as individual and isolated values.

III.2 Preparations for Metrological Investigations

The following paragraphs outline the preconditions generally required for determining receiver parameters by using the measuring procedures described. Special requirements will be described in the respective section.

As a general rule, testing should be performed under nominal conditions, which include especially:

(a) That the testing of a specimen is performed at the *specified operating* voltage.
(b) That the source impedance of the test signal used conforms to the impedance specified for testing the specimen interface, that is, is *nominally matched* (Section III.3). This applies particularly to procedures performed on the RF interface, that is, at the antenna socket. If necessary, a matching pad must be inserted (Fig. III.1). The attenuation figure of the pad has to be taken into account.
(c) That the AF output is terminated with the *nominal load*, having a load impedance as specified by the manufacturer. (Usually $4\,\Omega$, $8\,\Omega$ or $16\,\Omega$ for loudspeaker connections and $600\,\Omega$ for headphones. If only a meter with a high-resistance input is available, the condition can be easily met by inserting an ohmic resistor with the correct power rating in parallel.)
(d) That the reference output power (of limited distortion) at the AF output is adjusted with the volume control according to the manufacturer's specifications. The *reference*

Figure III.1 Example of a bidirectional matching pad allowing systems with a characteristic impedance of $50\,\Omega$ and $75\,\Omega$ to be matched up to $2.7\,\text{GHz}$. The ports must be connected to each other with the same characteristic impedance. Most test signal generators have a source impedance of $50\,\Omega$. Unidirectional matching pads are entirely suitable for their modification to feed $75\,\Omega$ systems and offer the advantage of low attenuation. The output voltage indicated at the test signal generator (the source) can then be used for the $75\,\Omega$ system without any further correction. (Company photograph of Rohde&Schwarz.)

output power can be determined easily by measuring the voltage across the load according to the formula

$$P_{\text{ref}} = \frac{V_{\text{nom}}^2}{R_{\text{nom}}}$$ (III.1)

where
P_{ref} = reference output power at the AF output, in W
V_{nom} = effective voltage across the nominal load, in V
R_{nom} = nominal load, in Ω.

(e) That all systems for *noise reduction* are *deactivated*.

Nominal modulation usually refers to a modulation frequency of 1 kHz. The name 'standard reference frequency' is also found in the professional literature. The class A3E emission of double sideband amplitude modulation with full carrier uses a modulation depth of 30%. With class F3E emission, that is frequency modulation, the peak frequency swing is generally one-fifth of the channel spacing intended for the respective operating mode, with a nominal frequency swing of 60%. (For conventional analog FM voice radio with 25 kHz channel spacing a frequency deviation of

$$\frac{25 \text{ kHz}}{5} \cdot 0.6 = 3 \text{ kHz}$$

is to be used for the modulated useful signal to be measured.) Emission class A1A, Morse telegraphy with keyed carrier and class J3E emissions for single-sideband modulation with suppressed carrier utilize an unmodulated carrier with a frequency offset against the receive frequency so that the 1 kHz AF signal is demodulated. The sideband in which the measurement is performed should be taken into account.

Especially when measuring the absolute level in higher frequency ranges it is important to consider the attenuation figure of the feeder cable used in the measuring setup. With multi-signal measurements the use of cables of identical electrical length and the same manufacturing type is recommended.

A SINAD meter is not always available. As long as it is not necessary to perform the measurements according to certain specifications, the determination of $(S+N)/N$ (Section III.4.8) can usually be substituted for the SINAD measurement with sufficient accuracy.

III.2.1 *The Special Case of Correlative Noise Suppression*

Some modern radio receivers feature correlative noise suppression methods with dynamic matching of the bandwidth to the useful signal without any operator action (and probably cannot be switched off by the operator). This can cause severe measuring errors in the classical analog single-tone measuring procedure! Such cases require either the complete deactivation of the noise elimination system or, to 'outsmart' the system, the measurement of signals having the same bandwidth as the emission class of interest in order to bypass

Figure III.2 The IFR39xx from Aeroflex belongs to a measuring station family of the latest generation. Owing to its modular design it can be configured for standardized TETRA receiver measurements. The unit also provides options for testing the basic parameters of radio receivers for AM and FM demodulation. (Company photograph of Aeroflex.)

the automatic bandwidths reduction of the noise elimination system. (For measurements in the J3E receive path it is helpful to use an FM-modulated signal.) However, the basic measuring procedure, by analogy to the methods described below, remains the same. In comparative tests performed on purely analog radio receivers, the equivalence of the measuring results must be verified in advance! If the circuitry of a test specimen is not fully known, the receiver noise figure can be estimated approximately by means of the minimum discernible signal (Section III.4.5) and the receive bandwidth (Section III.6.1), applying Equation (III.11). If the resulting value is clearly below the noise figure to be expected from a state-of-the-art device or even negative, it must be assumed that the receive paths tested make use of such a noise eliminating method.

III.2.2   The Special Case of Digital Radio Standards

The fields of digital mobile radio and digital trunked radio systems are governed by a number of relevant standards stipulating the test procedures for radio receivers. The guidelines are mandatory for such measurements. (A detailed description of the limiting conditions which apply for the meaningful comparison of receiving sensitivities of different radio standards can be found in [1], together with recommendations for suitable conversions.) The individual receiver parameters and their resulting effects also apply! Only the testing procedures are different. Testing is often carried out by fully automated communication analyzers or radio test sets (Fig. III.3). While for analog transmission systems the (demodulated) useful signal-to-interference ratio is an important quality criterion, in digital

Figure III.3   The 4400 Mobile-Phone Tester from Aeroflex is a cost-effective multi-standard test platform for mass-market terminal equipment (GSM, HSCSD, GPRS, EDGE, CDMA2000, 1xEV-DO, WCDMA, TD-SCDMA). Standardized basic measurements on receiver modules are performed largely automatically via the air interface in compliance with the applicable standard. The specimen (mobile terminal equipment) is simply placed into the test well and the necessary connections are made automatically. (Company photograph of Aeroflex.)

systems it is the bit error rate (Section III.4). The test signal simulating the useful signal is modulated by a data generator.

The signal-to-noise ratio (Section III.4.8) and the noise bandwidth (Section III.4.4) equivalent to the receive bandwidth determine the maximum possible transmission rate (also called transmission capacity or channel capacity) [2] according to the formula

$$C \approx 3,32 \text{ bit} \cdot B_N \cdot \frac{(S+N)/N}{10} \tag{III.2}$$

where

C = transmission capacity, in bits/s

B_N = noise bandwidth equivalent to the used receive bandwidth, in Hz

$(S+N)/N$ = (signal plus noise)-to-noise ratio, in dB

This is the highest average number of binary characters that can be transmitted per unit time. A transfer rate even close to the limit requires extensive coding efforts for the transmission of large data quantities. One distinguishes between:

(a) *source encoding*, which ensures that the information to be transmitted is described by as few basic symbols or binary characters per time unit as possible; and
(b) *channel encoding*, which has the task of re-coding the source-encoded code words to generate a signal suitable for the transmitting channel. The main purpose is the

detection and correction of transmission errors by means of a fault-tolerant or fault-correcting code. This requires code words (data packages) which differ in as many symbols as possible. The information rate is lower than the data rate since, besides the information content, other data, like fault detection data, have to be transported.

Data encoding (especially when large data volumes are concerned) has the disadvantage that the data package must be received as a whole and the information becomes available only afterwards [3]. The latency period permitted must not exceed a few characters which of course entails restrictions for encoding.

III.3 Receiver Input Matching and Input Impedance

The receive power available at the receiver input is the signal extracted from the wave field impinging on the effective antenna area. It is fed to the receiver via a feeder cable (coaxial cable or two-wire line). If the impedance at the receiving antenna, the characteristic impedance of the feeder, and the receiver input impedance are different from each other, then part of the receiving power supplied is reflected and therefore not available to the radio receiver for further processing (Fig. III.4). The heterodyning effect of the forward wave and reflected wave produces standing waves along the cable. The distances between the maxima and minima correspond to half the wave length (taking into account the velocity factor of the feed line, which relates to the vacuum conditions. If the receiver is disconnected and the feed line open or short-circuited, the total power coming from the receiving antenna would be reflected. This phenomenon of a non-terminated or short-circuited cable is called 'total reflection'.) In an ideal case, however, the impedance situation is

$$R_{RX} = Z_0 = R_{ant} \tag{III.3}$$

where
R_{RX} = receiver input resistance, in Ω
Z_0 = characteristic impedance of the feeder, in Ω
R_{ant} = feed point impedance of the use receiving antenna, in Ω

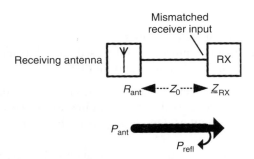

Figure III.4 Due to the mismatch between the receiving antenna and receiver input impedance $\underline{Z}_{RX}$ a part of the power P_{ant} coming from the antenna is reflected, shown in the signal flow diagram as P_{refl}.

so that correct *power matching* (also called impedance matching or common matching) ensures that no reflections occur. This results in the maximum power draw from the source (in this example the receiving antenna). The *receiver input resistance* is an ohmic resistance and, in this case, equal to the characteristic impedance of the feeder (usually $50\,\Omega$; it is therefore called a $50\,\Omega$ system).

In practice, however, this situation does not occur, especially not over a wide frequency range. In addition to the ohmic component, which differs to some degree from that of the characteristic impedance, an inductive component or a capacitive component contribute to the total *receiver input impedance*.

$$\underline{Z}_{RX} = R_{RX} \pm j X_{RX} \qquad (III.4)$$

where
$\underline{Z}_{RX}=$ complex receiver input impedance, in Ω
$R_{RX} =$ ohmic component of the receiver input impedance, in Ω
$X_{RX} =$ reactive component of the receiver input impedance, in Ω

The degree of *mismatch* is given by the ratio of the receiver input impedance and characteristic impedance and is expressed as ripple factor or standing wave ratio (SWR):

$$SWR_{RX} = \frac{1 + \sqrt{\dfrac{(R_{RX} - Z_0)^2 + X_{RX}^2}{(R_{RX} + Z_0)^2 + X_{RX}^2}}}{1 - \sqrt{\dfrac{(R_{RX} - Z_0)^2 + X_{RX}^2}{(R_{RX} + Z_0)^2 + X_{RX}^2}}} \qquad (III.5)$$

where
$SWR_{RX} =$ standing wave ratio of the receiver input, dimensionless
$R_{RX} =$ ohmic component of the receiver input impedance, in Ω
$Z_0 =$ characteristic impedance of the feed line, in Ω
$X_{RX} =$ reactive component of the receiver input impedance, in Ω

An SWR value of 1.0 indicates perfect matching. With an increasing degree of mismatch the SWR approaches infinity (total reflection).

Poor matching of the receiver input causes the greatest problems when an external receive preamplifier is used in order to increase the sensitivity (Section III.4). The mismatch between the antenna and the input of the receive preamplifier, as well as the narrow-band (selective) termination by the receiver input with its reactive component, often cause oscillations. Furthermore, with accurate receive level measurements (Section III.18) the measuring uncertainty increases with increasing mismatch of the receiver input. The reflection at the radio receiver input ultimately results in level uncertainties of the measured input signal due to the heterodyning forward and reflected waves. When activating a built-in RF preamplifier or RF attenuator, the mismatch of the receiver input can change depending on the circuit design, but may be improved by the RF attenuator (Figs. III.5 and III.6). Specifications in data sheets refer to the nominal impedance (today typically $50\,\Omega$) and state the standing wave ratio for the useful (fundamental) frequency band or the entire frequency range.

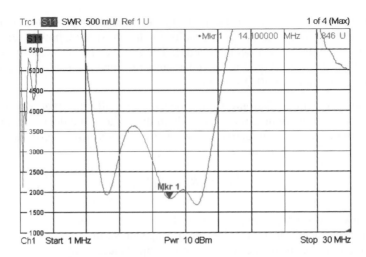

Figure III.5 Standing wave ratio for an HF radio receiver input tuned to 14.1 MHz, illustrated in the frequency range between 1 MHz and 30 MHz. The segment in which the standing wave ratio drops below 4 serves as an indication for the passband of the sub-octave input bandpass filter.

III.3.1 Measuring Impedance and Matching

For determining the receiver input impedance or testing the matching of the input of a receiver, a sensitive impedance measuring bridge, a sensitive active standing wave meter, a spectrum analyzer with tracking generator and an SWR measuring bridge or a vectorial network analyzer are used. The meter or its measuring head is connected directly to the

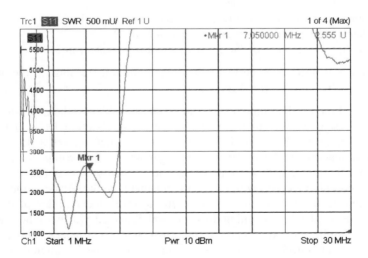

Figure III.6 Shift of the input standing wave ratio for the same HF radio receiver as in Figure III.5, caused by retuning to 7.05 MHz (instead of 14.1 MHz in Fig. III.5). This causes another sub-octave input bandpass filter to become active in the receive frequency band.

receiver input. The impedance is measured at the receive frequency to which the radio receiver is tuned. The RF level should be kept as low as possible in order to prevent overloading of the input stage.

III.3.2 Measuring Problems

The RF level of the measuring signal must be sufficiently low. Tests with an unknown specimen should start at 10 dB below the *IP*3 (Section III.9.8). If the RF frontend does not operate in the linear range, it can falsify the measuring result.

The stray radiation (Section III.17) of the receiver can result in measuring errors, especially with broadband instruments. Very often, this can be corrected only by using an operating meter that is as selective as possible for determining the matching. In such cases, reasonable results can be achieved only by using a spectrum analyzer with a standing wave ratio measuring bridge (or directional coupler) or a network analyzer with a small IF bandwidth. An acceptable input measurement is not possible on discrete frequencies with measurable receiver stray radiation. The better the selectivity of the meter used, the closer to the discrete interfering frequency point the measurement can be performed.

III.4 Sensitivity

The sensitivity of a radio receiver defines how well weak RF signals with a given modulation can be reproduced at the AF output and what the quality of the information content will be. This 'quality of the information content' *is always the result of a certain signal-to-noise ratio* (SNR) for a demodulated useful AF signal.

This can be expressed in various ways. While in the traditional transmissions types the direct indication of the SNR is commonly used, the character error rate (CER) or bit error rate (BER) is used in digital transmissions (Fig. III.7). The easiest method would

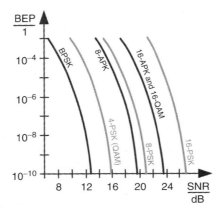

Figure III.7 Bit error probability (BEP) as a function of the SNR in various digital modulation methods. This indicates that the BEP changes by several orders of magnitude with a change of only a few dB in the signal-to-noise ratio [4].

be to count the number of actual errors. But this would not take either the transmission rate or the number of bits transmitted into account. According to [5] BER is the ratio of the number of bits, that are faultily demodulated at a certain RF input level, to the overall number of bits supplied. A typical presentation of the sensitivity would be:

-105 dBm for BER $= 10^{-3}$,

meaning that only one of 1,012 bits transmitted was faultily demodulated.

Another parameter that describes the sensitivity of a search receiver (Section II.4.2) used in the field of radio intelligence is the probability of detection. This characterizes the capability of the equipment to detect weak signals [6]. It refers for example to the input level for which a 99% probability of detecting the 'existing signal' can be guaranteed. In this context, the search speed (Section III.20) must be regarded as another important performance parameter for receivers used in radio intelligence and radio surveillance (Section II.4).

III.4.1 Limitations Set by Physics

The parameters having an actual effect on the 'quality of the information content' or the SNR of the AF output signal are

(a) the inherent noise of the radio receiver,
(b) the receive bandwidth (Section III.6.1) or equivalent noise bandwidth (Section III.4.4) of the receive path in its actual configuration, and
(c) the emission class (type of modulation) used.

The thermal noise sets the lower limit for detecting the lowest possible signals, even under ideal conditions. With power matching and an ambient temperature of 17 °C (62.5 °F, 290 K), the thermal noise is -174 dBm, measured with a bandwidth of one Hertz. Since noise is distributed in form and amplitude across a wide frequency range and cannot be captured as a signal component limited to a certain frequency, the intensity varies with the bandwidth observed (see Figs. III.13 and III.14). The wider the measured bandwidth, the higher is the (summed or averaged) noise power, which increases proportionally with the bandwidth. The noise power for a certain bandwidth must correctly be called the 'power spectral density' (PSD). (A rigorous description is given in [7] with the wording slightly adapted for the present context: It is not possible to extract a single frequency as a sinusoidal wave from the noise frequency spectrum, since noise is characterized by its unpredictability and randomness. It can be determined only as an average value within a certain bandwidth.) A fictitious bandwidth of one Hertz is used as a standard which, however, necessitates a conversion (Section III.4.4) to the bandwidth actually used. The thermal noise of -174 dBm is therefore correctly described as -174 dBm/Hz.

The actual influence of the ambient temperature can be neglected for practical reasons. When considering extreme conditions with temperature variations from -35 °C to $+45$ °C, the thermal noise power varies by no more than 1.26 dB over the entire temperature range. Corrections in regard to a value of -174 dBm/Hz are therefore unnecessary.

III.4.2 Noise Factor and Noise Figure

An ideal receiver, that is a receiver which is free of inherent noise, could detect extremely small signals with a receive bandwidth of one Hertz. Such a receiver would have a noise factor $F = 1$.

The *noise factor* (sometimes erroneously called noise figure) indicates how much higher the noise of the specimen is compared to that of an ohmic resistor of the same impedance at 290 K, measured over the same bandwidth. (The temperature of 290 K is the so-called reference temperature.) Equally valid is the statement that the noise factor is the ratio of the SNR at the input to the SNR at the output of the specimen when the noise source at the input is characterized by thermal noise at 290 K. This can be expressed as

$$F = \frac{\dfrac{P_{S\,in}}{P_{N\,in}}}{\dfrac{P_{S\,out}}{P_{N\,out}}} \tag{III.6}$$

where
    F = noise factor of the specimen, dimensionless
    $P_{S\,in}$ = power of the signal at the input of the specimen, in W
    $P_{N\,in}$ = thermal noise at 290 K at the input of the specimen, in W
    $P_{S\,out}$ = power of the signal at the output of the specimen, in W
    $P_{N\,out}$ = noise power at the output of the specimen, in W

(See Figs. III.8 and III.26.) To allow the easy use of decibel units it is only necessary to take the logarithm

$$F_{dB} = 10 \cdot \lg(F) \tag{III.7}$$

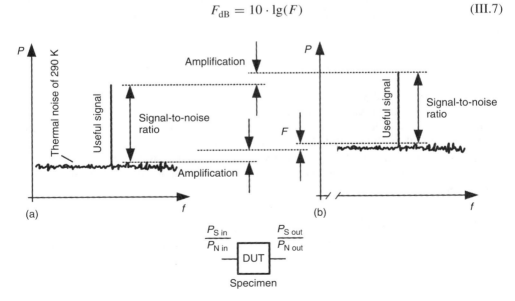

Figure III.8 Illustration of the noise factor for a specimen. In diagram (b) the signal is increased by the gain of the specimen. The signal-to-noise ratio is reduced by F of the specimen.

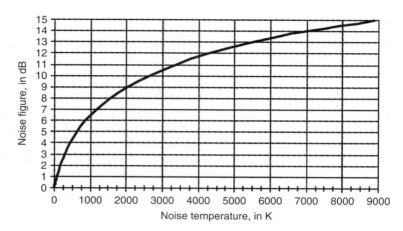

Figure III.9 Older publications sometimes quote the noise temperature in Kelvin instead of the noise figure; a direct conversion is possible. From today's point of view, the noise temperature is no longer of importance and is therefore not explained here in detail. A certain noise temperature can be assigned to every noise figure. The noise figure can be taken from this graph for noise temperatures up to 9,000 K.

where
F_{dB} = noise figure of the specimen, in dB
F = noise factor of the specimen, dimensionless

In accordance with DIN 5493-2 (where the German term used is 'Rauschmaß'), the correct name for the *logarithm of the noise factor* is *'noise figure'* (Figs III.9 and III.10). In the literature the term 'noise figure' is used as well. (Unfortunately, the two terms are sometimes incorrectly confused.) A noise factor of 16 corresponds to the noise figure of

$$F_{dB} = 10 \cdot \lg(16) = 12 \text{ dB}$$

Since every component generates some noise, it is not possible to build a receiver with $F = 1$ or $F_{dB} = 0$ dB. Casually speaking, the *receiver noise figure* indicates by how many dB the receiver under test is less sensitive than an ideal receiver, provided that the same receive bandwidth and demodulation type is used in both receivers! This value requires no additional information (like the receive bandwidth) and still – or precisely for this reason – allows rather objective and, most importantly, simple comparison of data from different sources.

The noise figure of a receiver is determined primarily by the noise figures and amplifications/attenuations of the upstream circuits (see also Section V.1) of the receive path, like the input selector, RF amplifier, first mixer, and to a certain amount the first IF stage.

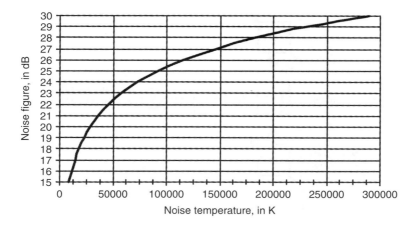

Figure III.10 Relation between noise figures between 15 dB and 30 dB and the corresponding noise temperatures (see also Fig. III.9). As can be seen, with high noise levels the noise temperature is difficult to use and of little practical benefit in regard to high resolution.

III.4.3 Measuring the Noise Figure

The setup for measuring the noise figure of a receiver using class A1A, A3E and J3E emissions is shown in Figure III.11.

Measuring procedure:

1. Tune the receiver to the frequency range to be tested.
2. Connect the true rms voltmeter or AF level meter to the AF output of the receiver.
3. Terminate the antenna socket using a dummy antenna, measure the AF output level and note the reading.
4. Replace the dummy antenna with a noise generator and increase its output level P_1 until the entire AF output level has increased by 3 dB or the AF output voltage reaches 1.41 times the original value.
5. Read the receiver noise figure directly from the calibrated noise generator.

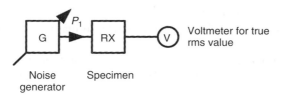

Figure III.11 Measuring arrangement for determining the noise figure of a receiver for demodulating class A1A, A3E, and J3E emissions.

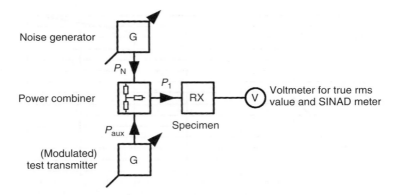

Figure III.12   Test measuring arrangement for the determination of the noise figure for a receiver using class F3E emissions.

If the noise generator shows the noise factor only, the reading can be converted into the noise figure with Equation (III.7).

The measuring arrangement for determining the noise figure of a receiver of emission class F3E is shown in Figure III.12. (The same setup should be used for class A3E emissions in a specimen with an AM envelope curve demodulator without biased diodes. However, today this is almost exclusively the case with older receivers.)

Measuring procedure:

1. Tune the receiver to the frequency range to be tested.
2. Connect the rms voltmeter or AF level meter and SINAD meter to the AF output of the receiver.
3. Terminate the input of the power combiner to which the noise generator will be connected later. Measure and note the AF output level.
4. Tune the test transmitter to the receive frequency. Modulate with nominal modulation (Section III.2) and increase the RF level P_{aux} until a linear relationship between P_{aux} and 10 dB to 16 dB SINAD is obtained at the AF output. (With F3E a linear relationship does not exist below the so-called FM threshold. The same is true for A3E until exceeding the threshold voltage of the diodes in the envelope curve demodulators, provided that these are not biased.)
5. Deactivate the modulation of the test transmitter, so that only the unmodulated carrier P_{aux} is available.
6. Replace the dummy antenna with a noise generator and increase its output level until either the entire AF output level is 3 dB higher or the AF output voltage reaches 1.41 times the original value.
7. Read the P_N value from the calibrated noise generator and note the reading.

If the noise generator shows the noise factor only, the reading can be converted into the noise figure with Equation (III.7). The noise factor of the receiver is the P_N value corrected by the attenuation figure of the power summation stage.

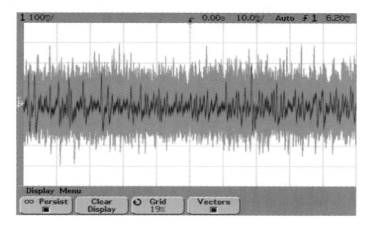

Figure III.13 Oscillogram of the noise at the AF output of a radio receiver. The IF filter inserted in the receive path has a measured bandwidth of 2,790 Hz at −6 dB with a shape factor of 1.7. (The black curve in the diagram represents the momentary output voltage at the time of observation, while the light gray area shows the maximum voltages occurring during the observation period of 10 s.)

III.4.4   Equivalent Noise Bandwidth

Since different receive bandwidths are used, every modulation type requires a certain minimum bandwidth and many radio receivers have no means of selecting the receive bandwidth required, the absolute value of the signal applied at the antenna socket for a defined sensitivity is of interest. As outlined in Section III.4.1 the noise level varies with the bandwidth (Figs. III.13 and III.14). This must also be true for the smallest signal still detectable in a certain receive bandwidth.

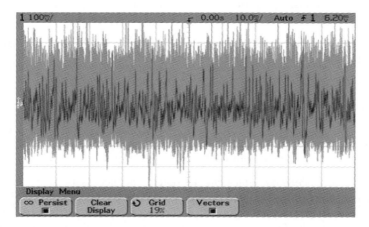

Figure III.14 Compared with Figure III.13 the noise level clearly increases by simply switching to a wider IF filter of 7,330 Hz at −6 dB bandwidth (shape factor 2.7). (The black curve in the diagram represents the momentary output voltage at the time of observation, while the light gray area shows the maximum voltages occurring during the observation period of 10 s.)

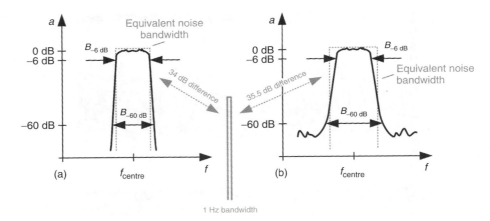

1 Hz bandwidth

Figure III.15   With the same −6 dB bandwidth the filter shown in (a) has a lower equivalent noise bandwidth than the filter in (b) with its relatively flat edge steepness and poorer far-off selection.

Even with an identical receive bandwidth (−6 dB bandwidth) a filter or receive path with a low shape factor (Section III.6.1) would allow less noise to pass than a device with a poor shape factor. The *equivalent noise bandwidth* has been introduced especially for such cases. This is determined by converting the area under the selection curve of the passband region to a rectangle having the same area and a height that corresponds to the maximum height of the real curve (Fig. III.15). The width of the 'rectangular filter' must be such that the same noise power passes as with the real filter [8].

This is the reason why the term 'receive bandwidth' is used more often than the IF bandwidth. It is not necessarily the IF bandwidth that has the strongest influence on the noise bandwidth; see, for example, the limitation of the AF frequency response (Section III.13.1). Decisive for the noise bandwidth is the narrowest bandwidth used in the receive path.

The increase of the noise level as compared to the 1 Hz bandwidth (Fig. III.15) can be determined from the equivalent noise bandwidth with

$$B_{\mathrm{dB\,N}} = 10 \cdot \lg \left(\frac{B_{\mathrm{N}}}{1\,\mathrm{Hz}} \right) \tag{III.8}$$

where
$B_{\mathrm{dB\,N}} =$ equivalent noise bandwidth of the receive bandwidth used, in dBHz
$B_{\mathrm{N}} =$ equivalent noise bandwidth of the receive bandwidth used, in Hz

The filters used in *receiver engineering* to obtain the necessary selectivity have such steep edges that converting the areas under the curve by integration is of no practical significance. For shape factors below 2, sufficient results are achieved by taking the −6 dB receive bandwidth for the equivalent noise bandwidth. With shape factors above 2, a correction is advisable [9], [10] using

$$B_{\mathrm{N}} \approx B_{-6\,\mathrm{dB}} \cdot 0.5 \cdot SF \tag{III.9}$$

where

B_N = equivalent noise bandwidth of the receive bandwidth used, in Hz

$B_{-6\,dB}$ = receive bandwidth ($-6\,dB$ bandwidth) of the receive path, in Hz

SF = shape factor describing the near selectivity of the receive path, dimensionless

In practice, one should remember that switching to a receive bandwidth of double the width causes an increase in the noise level by approximately 3 dB, while the SNR decreases by the same amount. This makes the noise component sound louder and fuller.

III.4.5 Minimum Discernible Signal

With class A1A and J3E emissions the signal power required at the antenna socket to obtain an increase of 3 dB at the AF output is called the minimum discernible signal (power) and can be calculated by a simple addition:

$$P_{MDS}(B_{-6\,dB}) = -174 \text{ dBm/Hz} + F_{dB} + B_{dB\,N} \qquad \text{(III.10)}$$

where

$P_{MDS}(B_{-6\,dB})$ = minimum discernible signal of the receiver with the receive
bandwidth ($B_{-6\,dB}$), in dBm

F_{dB} = noise figure of the receiver, in dB

$B_{dB\,N}$ = equivalent noise bandwidth for the used receive bandwidth, in dBHz

The minimum discernible signal (MDS) is also called the equivalent input noise power or the noise floor. The minimum discernible signal has a signal power equal to the noise power of the receive path. Due to the addition of the injected power and the inherent noise power, the level increases by 3 dB.

This is illustrated in Figure III.16. In this graph white noise with an average level of -55 dBm (from a noise generator) is added to a sinusoidal signal of the same level. The latter clearly exceeds the noise by 2.2 dB in the centre of the graph. The same happens in a radio receiver throughout the linear receiver circuitry. An experienced CW operator may be able to just barely perceive such an incoming signal at the output of the radio receiver. The difference of 0.8 dB is caused by the fact that by averaging the signal over various cycles the noise intensity continuously varies and becomes slightly lower during the collection period. The current professional literature and especially the application booklets from Agilent Technologies quote both values. All considerations below are based on the classical 3 dB increase, since this is the most common case in real procedures in most known laboratories and is specified in most testing instructions.

The equivalent noise bandwidth of an A1A receive path having a noise figure of 15 dB and a receive bandwidth of 500 Hz and a shape factor of 2.7 is given by Equation (III.9):

$$B_N \approx 500 \text{ Hz} \cdot 0.5 \cdot 2.7 \approx 675 \text{ Hz}$$

Then according to Equation (III.8):

$$B_{dB\,N} = 10 \cdot \lg\left(\frac{675 \text{ Hz}}{1 \text{ Hz}}\right) = 28.3 \text{ dBHz}$$

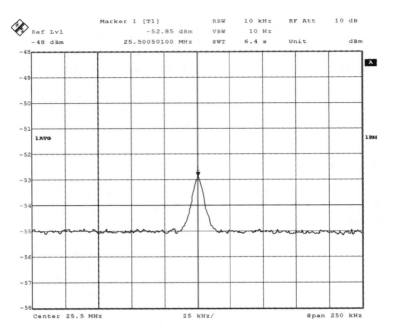

Figure III.16 Summation of the averaged noise and a injected sinus signal of the same signal level. The $(S+N)/N$ ratio amounts to 2.2 dB.

This leads to a minimum discernible signal power of

$$P_{\mathrm{MDS}}(B_{-6\,\mathrm{dB}} = 500\ \mathrm{Hz}) = -174\ \mathrm{dBm/Hz} + 15\ \mathrm{dB} + 28.3\ \mathrm{dBHz} = -130.7\ \mathrm{dBm}$$

III.4.6 Measuring the Minimum Discernible Signal

The setup for measuring the minimum discernible signal is shown in Figure III.17. It is common practice to perform separate measurements for class A1A and J3E emissions only as outlined below.

Measuring procedure:

1. Tune the receiver to the frequency range to be tested.
2. Connect the true rms voltmeter or AF level meter to the AF output of the receiver.

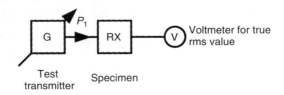

Figure III.17 Measuring arrangement for determining the minimum discernible signal.

3. Terminate the antenna socket using a dummy antenna.
4. Note the AF output level reading.
5. Tune the test transmitter to the frequency offset relative to the receive frequency, in which a 1 kHz tone is expected after demodulation. (Observe the upper and lower sidebands.)
6. Increase the output level P_1 of the test transmitter until the entire AF output level has increased by 3 dB or the AF output voltage reaches 1.41 times the original value.
7. Note the reading of P_1.

P_1 represents the minimum signal power discernible by the receiver in the receive bandwidth used.

When no noise generator is available for determining the receiver noise figure, the procedure can be reversed; from Equation (III.10) one then obtains

$$F_{dB} = P_{MDS}(B_{-6\,dB}) - B_{dB\,N} + 174 \text{ dBm/Hz} \qquad \text{(III.11)}$$

where
$$F_{dB} = \text{receiver noise figure, in dB}$$
$$P_{MDS}(B_{-6\,dB}) = \text{minimum discernible signal of the receiver with the receive bandwidth}$$
$$(B_{-6\,dB}), \text{ in dBm}$$
$$B_{dB\,N} = \text{equivalent noise bandwidth for the receiver bandwidth used, in dBHz}$$

Thus, from the *measured* minimum discernible signal of -130.7 dBm the receiver noise figure can be determined for the A1A receive path used as an example in Section III.4.5 using the relationship

$$F_{dB} = -130.7 \text{ dBm} - 28.3 \text{ dBHz} + 174 \text{ dBm/Hz} = 15 \text{ dB}$$

III.4.7   Input Noise Voltage

For historic reasons the receiver sensitivity is often expressed as a voltage, since in test transmitters the display of the signal strength was nearly always calibrated as a voltage and RF millivoltmeters were used to measure RF signal levels. Modern test equipment usually indicates signal levels in dBm. In matched systems, the conversion of one unit to the other is possible (Section V.7.1), and although the respective calculation is relatively simple the unit millivolt is persistently used for sensitivity indication. It is necessary to distinguish between the electromotive force (EMF or e.m.f.) and the voltage at the antenna terminals.

EMF is the open-circuit voltage or source voltage. This is the voltage actually generated at the source (test transmitter) and includes voltage losses across the internal resistance of all the loads. It can be measured with a high-resistance probe at the output socket. The formula symbol is V_{EMF}. Specifying data in this way may well be based on the assumption that comparing sensitivities is possible even between mismatched receiver inputs (Section III.3), since the EMF is independent of the effective test transmitter load. (One is reminded of the introduction of the 50 Ohm technology which replaced the earlier

60 Ohm technology and of the 75 Ohm technology now generally used in broadcasting technology.)

In contrast, the terminal voltage is the real voltage at the antenna socket within a matched system. Under this assumption the terminal voltage has half the value of the EMF (difference of 6 dB).

According to [6] and [11] the source voltage (EMF) equivalent to the minimum discernible signal is called the *input noise voltage*. For the example of the A1A receive path with its minimum discernible signal of -130.7 dBm (described in Section III.4.5), the input noise voltage is 0.13 µV EMF.

III.4.8   Signal-to-Interference Ratio (SIR) and Operational Sensitivity (S+N)/N, SINAD

The relation between the noise and a signal of the same level can be expressed as a signal-to-noise ratio (SNR) of 0 dB, since both components have the same signal power. *S/N* is another common abbreviation for signal-to-noise ratio.

Only if additional information is available on the expected signal is it possible to make use of correlation techniques, even with negative SNR values ('correlation' means 'related to each other'). Amplifying the signal brings no benefit, since the noise floor is also amplified and, due to the noise contribution of the amplifier, the SNR at the output is even lower [7]. For practical reasons it is almost impossible to determine the SNR by first measuring the signal and then the noise component (see Section III.4.5 and Fig. III.16). Noise cannot simply be turned off or eliminated, so that the signal is never isolated and must be measured together with the noise in the form of $S + N$. The term '(signal plus noise)-to-noise' $(S + N)/N$ is therefore only an auxiliary construct which, in the real world, is always present. The higher the signal-to-noise ratio, the smaller is the actual difference. With an increasing signal power, the influence by the noise power, the level of which remains the same, on the overall power measured becomes smaller. With an $(S + N)/N$ of 7 dB, measured within a moderate bandwidth, the difference relative to the SNR is less than 1 dB.

Demodulation often produces harmonics of the AF signal. These too can compromise the intelligibility or the 'quality of the information content'. Simultaneous evaluation of the noise and the harmonic content (Section III.13.3) provides the value known as SINAD, that is, $(S + N + D)/(N + D)$. The abbreviation stands for '(signal plus noise plus distortion)-to-(noise plus distortion)'. This is calculated from the formula

$$\text{SINAD} = 20 \cdot \lg \left(\frac{V_{\text{tot}}}{\sqrt{V_N^2 + V_{2.\,\text{HW}}^2 + V_{3.\,\text{HW}}^2 + V_{4.\,\text{HW}}^2 + \cdots + V_{n.\,\text{HW}}^2}} \right) \qquad \text{(III.12)}$$

where
SINAD = (signal plus noise plus distortion)-to-(noise plus distortion), in dB
V_{tot} = effective value of the total signal, in V
V_N = effective value of the noise component, in V

$V_{2.\,HW}$ = effective value of the 2nd harmonic, in V
$V_{3.\,HW}$ = effective value of the 3rd harmonic, in V
$V_{4.\,HW}$ = effective value of the 4th harmonic, in V
$V_{n.\,HW}$ = effective value of the nth harmonic, in V

Splitting the signal into its spectral components by converting it from the time domain to the frequency domain is illustrated in Figure III.93 with the signal components according to Equation (III.12).

Measurements of the receiver sensitivity using SINAD meters are very efficient. The automated insertion and removal of the demodulated useful signal shows both the total signal scenario, on the one hand, and the noise together with the AF harmonics alone on the other hand (Figs. III.18 and III.19). The SINAD figure in dB is indicated directly. In order to avoid measuring errors it is only necessary to consider that the useful signal frequency is equal to the centre frequency of the notch filter in the SINAD meter. Measuring signals with a low harmonic distortion after processing in the radio receiver, the two values $(S+N)/N$ and SINAD are equal.

With class A1A emission keying always produces the full power. This is not the case with other emission classes (like the modulated power in A3E or J3E, which varies with the envelope curve). Since many emission classes are of a more complex nature and, after demodulation, deliver not only a single tone but an entire voice band, a high SNR is required for high sound quality. The parameter *operational sensitivity* (sometimes called the reference sensitivity) of a radio receiver refers to a certain $(S+N)/N$ or SINAD at the AF output (by reference output power). Values of 10 dB, 12 dB or 20 dB are common and often stipulated in test specifications (Figs. III.20, III.21 and III.22). An SNR of 33 dB represents the minimum quality requirement for broadcast receivers.

By inserting weighting filters in front of the measuring device for evaluating the SNR at the AF output, it is possible to evaluate the measured results with respect to physiological hearing ability. At the same time this causes a limitation of the AF frequency response, so that measurements with inserted weighting filters result in subjectively higher receiver sensitivities. Some test specifications require such measures. According to the standard P53A of CCITT (Comité Consultatif International Télégraphique et Téléphonique), radio

Figure III.18   Close-up view of the scale of a SINAD meter. For tuning purposes, analog displays often enable a more efficient workflow.

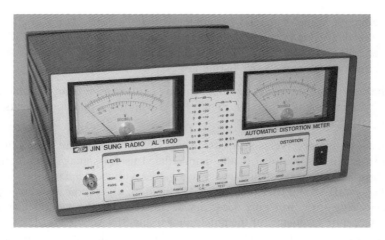

Figure III.19   Instrument AL1500 from JSR combining a SINAD meter and an AF millivoltmeter (with level indicator) and distortion measuring bridge.

telephone engineering often demands a bandpass filter of standardized passband characteristics (generally called the CCITT filter) (Fig. III.23). For the signal-to-noise ratio it would be technically correct to distinguish between the *unweighted (noise) voltage* or *unweighted signal-to-noise ratio* (without a weighting filter inserted) and the *improved noise voltage* or *weighted signal-to-noise ratio*. If the AF output signal of a radio receiver is not expressly related to acoustic reception but used for further processing like in decoders, measurements with a weighting filter are not very meaningful.

Following Section III.4.7, there are several ways to express operational sensitivity. Table III.1 lists commonly used specifications for one and the same radio receiver with

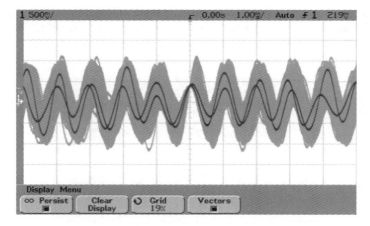

Figure III.20   Demodulated 1 kHz signal with 10 dB SINAD at the AF output of a radio receiver. (The black curves represent the output voltage measured at the time of observation. The light gray area shows the variations due to noise influences over an observation time of 10 s.)

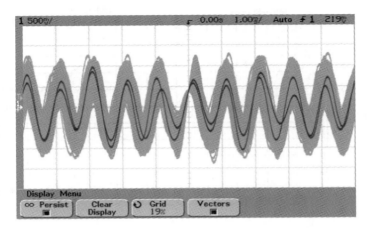

Figure III.21 Demodulated 1 kHz signal with 12 dB SINAD at the AF output of a radio receiver. Compared with Figure III.20 the amplitudes are higher. (The black curves represent the output voltage measured at the time of observation. The light gray area shows the variations due to noise influences over an observation time of 10 s.)

a 50 Ω input and low-distortion signal processing as an example. All of these have the same meaning.

Whenever receiver sensitivity is actually determined by the inherent noise of the radio receiver, the term signal-to-noise ratio or useful signal-to-noise ratio is used. Where interference signals (e.g., with reciprocal mixing (Section III.7) or intermodulation

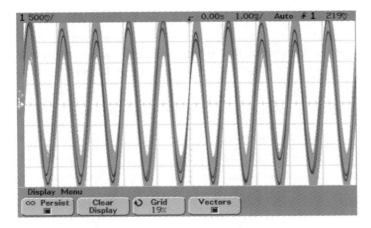

Figure III.22 Demodulated 1 kHz signal with 20 dB SINAD at the AF output of a radio receiver. While the unpredictable short-term variations due to changing noise levels correspond approximately to the effective voltages in Figures III.20 and III.21, the absolute signal amplitude is clearly more dominant. (The black curves represent the output voltage measured at the time of observation. The light gray area shows the variations due to noise influences over an observation time of 10 s.)

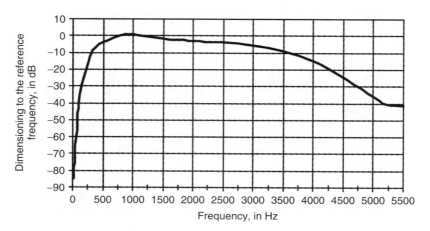

Figure III.23 Selection characteristic of a standardized weighting filter according to CCITT (Comité Consultatif International Télégraphique et Téléphonique) mainly used in radio telephony engineering. The reference frequency is 800 Hz.

However, the CCITT standard is outdated, although the name is still used for one of the technical committees of ITU (International Telecommunications Union). This committee has assumed the name ITU-T (ITU Telecommunication Standardization Sector) and is the organizational unit of the ITU for developing technical standards and recommendations for telecommunication applications [12].

(Section III.9)) affect the useful signal, the expression *signal-to-interference ratio* (SIR) or *useful signal-to-interference ratio* should be used.

III.4.9   De-emphasis

With VHF FM broadcasting and some other (voice) communication networks based on frequency or phase modulation a further improvement is achieved by intentionally provoking a non-linear transfer frequency response. For this purpose the higher modulation

Table III.1   Synonymous specifications for the operational sensitivity of a radio receiver

Emission class	Remarks	Operational sensitivity
F3E (FM narrow)	No weighting filter	0.13 µV at 50 Ω for 12 dB $(S+N)/N$
F3E (FM narrow)	No weighting filter	0.13 µV at 50 Ω for 12 dB SINAD
F3E (FM narrow)	No weighting filter	0.26 µV EMF for 12 dB $(S+N)/N$
F3E (FM narrow)	No weighting filter	0.26 µV EMF for 12 dB SINAD
F3E (FM narrow)	No weighting filter	−17.7 dBµV at 50 Ω for 12 dB $(S+N)/N$
F3E (FM narrow)	No weighting filter	−17.7 dBµV at 50 Ω for 12 dB SINAD
F3E (FM narrow)	No weighting filter	−11.7 dBµV EMF for 12 dB $(S+N)/N$
F3E (FM narrow)	No weighting filter	−11.7 dBµV EMF for 12 dB SINAD
F3E (FM narrow)	No weighting filter	−124.7 dBm for 12 dB $(S+N)/N$
F3E (FM narrow)	No weighting filter	−124.7 dBm for 12 dB SINAD

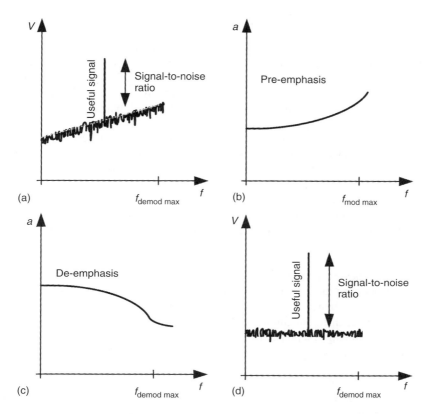

Figure III.24 Graphic illustration of SNR improvement by pre- and de-emphasis. (a) Shows a demodulated AF signal spectrum with an almost linear modulation frequency response for the transmitter and receiver. (b) Shows the frequency response of a transmitter using pre-emphasis. (c) Illustrates how the frequency response of the receive path using de-emphasis removes the emphasis on higher frequencies. (d) Clearly shows a higher SNR resulting from the demodulation of this signal with de-emphasis.

frequencies are increased in a defined manner at the transmitter. This method is called *pre-emphasis* (here in the sense of greater emphasis or 'accenting').

In the receive path the now stronger higher frequencies are reduced together with the higher frequency noise components by means of a low-pass filter having an oppositely directed frequency response [6]. This method is called *de-emphasis* (removing the emphasis). This results in an improvement of the signal-to-noise ratio (Fig. III.24). It is important that the networks influencing the frequency response for pre-emphasis and de-emphasis are compatible, in order to avoid an undesired amplitude frequency response [13]. The frequency-modulated signal altered in this way becomes a quasi phase-modulated signal.

When determining the sensitivity of receive paths using de-emphasis, the measurement in cases of purely F3E modulated RF input signals (from the test transmitter) provide sensitivity values that are too good [14].

III.4.10   Usable and Suitable Sensitivity

When receiving signals under operational conditions the actually usable sensitivity of a radio receiver strongly depends on the receive frequency. This is due to the fact that all signals of noise characteristics received from the antenna recombined to a noise voltage that is often higher than the thermal noise at the antenna feeding point impedance. This is called *external noise*. The primary sources (of radiation) are a combination of the following:

(a) *Atmospheric noise* – with low receive frequencies this is caused by lightning discharges worldwide (a high number of lightning events per second, especially in tropical areas), and with high receive frequencies it is caused by atmospheric gases and hydrometeors (collective term for all forms of condensed water present in the atmosphere).

(b) *Galactic and cosmic noise* – the noise of radioactive celestial bodies in the centre of the Milky Way and of the sun (solar noise) contained in (high) receive frequencies for which the ionosphere is permeable. The minimum is given by cosmic background radiation.

(c) *Earth noise* – noise signals originating from the earth or the earth's surface, since every absorbing body emits radiation depending on the temperature, surface roughness, air humidity, and frequency. For example, a concrete surface or a rocky mountain massif emits noise of a different intensity than grassland [15]. (A hypothetical ideal black body [16] has the highest possible absorption capability and emits the strongest thermal radiation power possible according to the laws of physics for the given temperature. The universal character of the thermal radiation emitted by a black body and the fact that at any given frequency no real body can emit a stronger radiation than a black body suggest that it is reasonable to describe the emission capability of a real body in relation to that of a black body. The ratio of the radiation intensity emitted by a real body to that from a black body of the same temperature is called the emissivity and can range between 0 and 1 [16]. The earth displays different colours and is not a black body, which explains the different noise intensities.)

(d) *Thermal noise or Johnson noise, Nyquist noise or thermal resistance noise* – this is caused by ohmic loss resistances due to the varying mean electron speed resulting from lattice vibrations transferred to the movement of charge carriers. (Resistance noise is by definition not an external noise, but is provided by real antennas at the antenna feeding point due to material losses. Its real contribution to the entire noise power yielded by the antenna is determined by the antenna efficiency.)

(e) *Man-made noise, technical noise or industrial noise* – this is the result of all kinds of electrical switching actions, sparks, discharges, pulses, and the like due to insufficient, defective or impossible electromagnetic shielding/interference suppression. (Sources can be spark discharges, collector effects, switching actions in the lighting system, phase-fired controls, etc.).

Received from far away via the air interface, signals have noise characteristics (Fig. III.25). Their accumulated level can, especially with low receive frequencies and up to the VHF range, be so strong that the weakest useful signal received can no longer be discriminated from the inherent noise of the receiver. In a *varied form* of the definition given for the noise factor and noise figure (Section III.4.2 and Fig. III.26), the external noise figure can

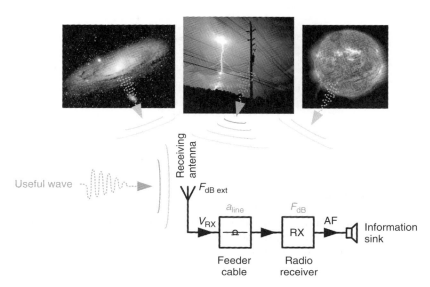

Figure III.25 Electromagnetic radiation from celestial bodies like the sun (at the right), lightning, and energy-carrying (high voltage) power lines contribute to external noise. The quality of useful signals can be severely affected or 'masked' entirely, despite a sufficiently low receiver noise figure.

be used as a universally applicable expression as the ratio of all the noise power received to the thermal noise power. According to [17] the *external noise figure* (Fig. III.27) defines how much *more or less* noise power in dB an antenna provides at 290 K than an ohmic resistor having the same resistance as the real component of the antenna input impedance. (The terms 'antenna noise figure', 'external noise figure' or 'noise temperature' can also be found in the technical literature (Figs. III.26 and III.27).) Table III.2 shows that the

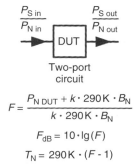

$$F = \frac{P_{N\,DUT} + k \cdot 290\,K \cdot B_N}{k \cdot 290\,K \cdot B_N}$$

$$F_{dB} = 10 \cdot \lg(F)$$

$$T_N = 290\,K \cdot (F - 1)$$

Figure III.26 Definition of the noise factor, noise figure, noise temperature in two-port circuits. Contrary to the external noise factor/figure (Fig. III.27) the available thermal noise power at the reference temperature has to be taken into account in the numerator. With this difference in the definition simple approaches to the receiving environment become possible, as shown in Table III.2. ($k = 1.38 \cdot 10^{-23}$ Ws/K; $P_{N\,DUT} = P_{N\,out} - G \cdot P_{N\,in}$, in W; B_N = equivalent noise bandwidth, in Hz)

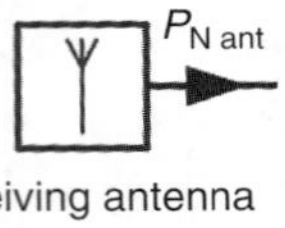

Receiving antenna

$$F_{\text{ext}} = \frac{P_{\text{N ant}}}{k \cdot 290\,\text{K} \cdot B_{\text{N}}}$$

$$F_{\text{dB ext}} = 10 \cdot \lg(F_{\text{ext}})$$

$$T_{\text{N ext}} = 290\,\text{K} \cdot F_{\text{ext}}$$

Figure III.27 Definition of the external noise factor, noise figure, noise temperature (see also Fig. III.26) at the receiving location by means of the available noise power $P_{\text{N ant}}$ of a loss-free antenna. With strongly focusing (cooled) receiving antennas of high efficiency lower received noise power is possible, like the available thermal noise (in the denominator) at the reference temperature. This explains negative external noise figures (e.g., of cosmic background radiation). ($k = 1.38 \cdot 10^{-23}$ Ws/K; $B_{\text{N}} =$ equivalent noise bandwidth, in Hz)

smallest signal discernible by a receiving system consisting of the receiving antenna and the connected receiver is determined by the external noise if the external noise figure is higher than the receiver noise figure. It is therefore desirable that

$$F_{\text{dB}} < F_{\text{dB ext}} \tag{III.13a}$$

where

F_{dB} = receiver noise figure, in dB
$F_{\text{dB ext}}$ = external noise figure, in dB

and

$$F_{\text{dB}} + a_{\text{line}} < F_{\text{dB ext}} \tag{III.13b}$$

where

F_{dB} = receiver noise figure, in dB
a_{line} = feeder line attenuation figure, in dB
$F_{\text{dB ext}}$ = external noise figure, in dB

The intensity of the external noise during receiver operation depends to a large degree on

- the receive frequency,
- the receiver location (quiet country setting or urban industrial environment),

Table III.2 Negative impact of external noise on the receiving system

Receiver noise figure	Effect on the sensitivity
$F_{\text{dB}} = F_{\text{dB ext}}$	Sensitivity performance is affected by $\sim$3 dB
$F_{\text{dB}} = F_{\text{dB ext}} - 6\,\text{dB}$	Sensitivity performance is reduced by $<$1 dB
$F_{\text{dB}} < F_{\text{dB ext}} - 10\,\text{dB}$	No relevant loss in sensitivity

- the operating time (time of day, time of year) and the (ionospheric) propagation conditions,
- the possible directional characteristic, the full width of half maximum (FWHM), the elevation of the radiation pattern of the receiving antenna used, and
- the antenna orientation.

A certain quantification is possible with the data and evaluations given in [17]. The data obtained in the course of extensive measuring campaigns throughout the world (Figs. III.28, III.29 and III.30) replace the often cited report 322 of the CCIR (Comité Consultatif International des Radio Communications). These data are intended to provide a guideline for design concepts. (If not stated otherwise, an omnidirectional radiation pattern is assumed for the antennas shown in the three figures. With directional antennas having a strong focusing effect the deviation of the atmospheric noise due to lightning can be expected to be about ±5 dB in the RF range, depending on the orientation and location of the antenna. The solid curves indicate the minima expected under real conditions.) Reference [18] describes the generation of the data set and its development, as well as the models used. There is wide room for interpretation, since a generally applicable more accurate characterization is hardly possible because of the lack of strict conditions and the discontinuities.

It can be said that in a receive range below 35 MHz the weakest discernible signals are (very) rarely defined by the sensitivity of a state-of-the-art receiver; that is, the requirement of Equation (III.13) is almost always satisfied. The requirements of a high sensitivity and an equally high intermodulation immunity (Section III.9.6) are diametrically opposed to each other. This means that unreasonably low receiver noise figures reduce the large-signal tolerance (Section III.12). The higher the receive frequency, the more likely that the lowest discernible signal is determined by the receiver sensitivity! Here, the advantages of

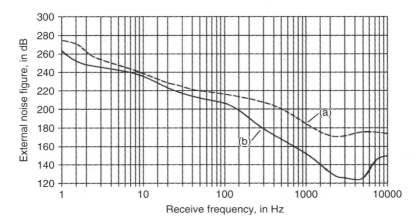

Figure III.28 External noise figure in the frequency range between 1 Hz and 10 kHz according to [17]. (a) Shows the maximum expected intensity that is seldom exceeded and (b) shows the minimum expected intensity below which values seldom occur. This is caused by atmospheric noise.

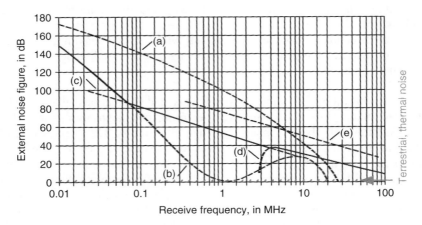

Figure III.29   External noise figure in the receive frequency range between 10 kHz and 100 MHz according to [17]. The intensity is broken down into contributions from (a) atmospheric sources (maximum, possibly higher in ∼0.5% of the time), (b) atmospheric sources (minimum, possibly higher in ∼99.5% of the time), (c) technical sources (quiet countryside), (d) galactic sources and (e) technical sources (busy industrial environment). The expected external noise with frequencies below 6 MHz is for example in 99% of the time between (a) and (b).

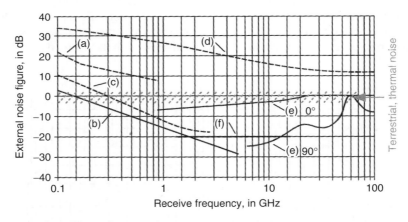

Figure III.30   External noise figure in the receive frequency range between 100 MHz and 100 GHz according to [17]. The intensities are broken down into contributions from (a) technical sources (busy industrial environment), (b) galactic sources, (c) the centre of the galaxy (detected by antennas having extremely narrow radiation lobes), (d) calm sun (antenna with 0.5° FWHM, oriented directly to the centre), (e) oxygen and water vapor (detected by antennas having extremely narrow radiation lobes with 0° and 90° elevation) and (f) cosmic background radiation as the absolute minimum with $F_{dB\ ext} = -20.3$ dB (corresponding to 2.7 K noise temperature).

a noise-optimized receiving system as described in Section V.1 are obvious. A *complete receiving system*, consisting of the antenna and a receiver connected via the feeder, reaches the *physical sensitivity limits* [19] according to Section III.4.5 when

- the external noise figure in the receive frequency range is clearly above the receiver noise figure plus the feeder attenuation figure (which is considered a part of the receiver noise figure for the remainder of this section); and
- the external noise figure (cosmic, atmospheric, man-made noise) is significantly higher than the inherent noise of the antenna (caused by the thermal noise of the loss resistances and, in case of active antennas, the amplifier noise).

According to [18] the expected interference field strength due to the external noise can be determined approximately by

$$E_{\text{ext}}(B_{-6\,\text{dB}}) \approx F_{\text{dB ext}} + 20 \cdot \lg\left(\frac{f_{\text{op}}}{10^6\,\text{Hz}}\right) + B_{\text{dB N}} - 100.3\,\text{dB}(1/m) + G_{\text{dBi ant}} \quad (\text{III.14})$$

where

$E_{\text{ext}}(B_{-6\,\text{dB}})$ = interference field strength level of the external noise, depending on the receive bandwidth used $(B_{-6\,\text{dB}})$, in dB(μV/m)

$F_{\text{dB ext}}$ = external noise figure, in dB

f_{op} = operating frequency, in Hz

$B_{\text{dB N}}$ = equivalent noise bandwidth of the receive bandwidth, in dBHz

$G_{\text{dBi ant}}$ = antenna gain figure of the antenna used, in dBi

When receiving the 41 m broadcasting band with a short vertical omnidirectional antenna with a real antenna gain figure of about 1.8 dBi and a receive bandwidth of 9 kHz, an external noise figure of 34 dB is expected according to Figure III.29. The expected interfering field strength level is then given by

$$E_{\text{ext}}(B_{-6\,\text{dB}} = 9\,\text{kHz}) \approx 34\,\text{dB} + 20 \cdot \lg\left(\frac{7.325\,\text{MHz}}{1\,\text{MHz}}\right)$$
$$+ 39.5\,\text{dBHz} - 100.3\,\text{dB}(1/m) + 1.8\,\text{dBi} \approx -7.7\,\text{dB}(\mu\text{V/m})$$

or 0.4 μV/m, if expressed as a voltage. At locations with high environmental radiation (urban installations) the expected interfering field strength is up to 22 dB stronger, as suggested by curve e) of Figure III.29. For reception in the centre of the VHF broadcast band with a measuring bandwidth and thus a receive bandwidth of typically 120 kHz, this changes to

$$E_{\text{ext}}(B_{-6\,\text{dB}} = 120\,\text{kHz}) \approx 10\,\text{dB} + 20 \cdot \lg\left(\frac{97.75\,\text{MHz}}{1\,\text{MHz}}\right)$$
$$+ 50.8\,\text{dBHz} - 100.3\,\text{dB}(1/m) + 1.8\,\text{dBi} \approx 2.1\,\text{dB}(\mu\text{V/m})$$

or 1.3 μV/m, when expressed as a voltage. In environments with strong industrial noise, a field strength level of up to 17 dB(μV/m) must be assumed in the receive channel when using a bandwidth of 120 kHz. For the above calculations the noise bandwidth equivalent to the receive bandwidth used has been determined with Equation (III.8) under the assumption of low (good) shape factors.

III.4.10.1 Improving the Reception in Environments with Pronounced Man-Made Noise

Optimization of the signal-to-interference ratio in situations of poor receiving situation can often be achieved by using antennas that predominantly pick up the magnetic field components. These are called magnetic receiving antennas. Examples are magnetic dipoles in the form of frame antennas or ferrite antennas, consisting of a conductive loop that is short compared to the wavelength [20]. While the influence of atmospheric or galactic disturbances can barely be influenced (except by using directional antennas, which are usually difficult to handle), the situation is quite different with technical man-made noise. The sources of interference (household appliances, electric machines, phase-fired controls, etc.) have much smaller dimensions than the radiated wavelengths, and the connecting cables seldom have the optimum length of a multiple of one-quarter of the radiated wavelength. The interfering signal therefore propagates mainly in the form of electric and magnetic, quasi static, near fields and not in the form of 'true' free-space propagation in the far field. Near fields are characterized by the fact that their field strengths drop at least according to the square of the distance from the source of interference. Furthermore, many such sources generate mainly interference in the form of electric fields. The reason is that the power line has two conductors in close proximity as twisted pairs, so that with interfering currents of opposing phases the magnetic fields are cancelled out. The interfering common-mode voltage of the same phase on both conductors generates the electric near field. The result is that reception with magnetic antennas is characterized by less interference than with electric antennas, provided that the source of interference of the cable emitting interference is in close proximity to the receiving antenna [20]. Of course, this effect is much stronger in receive frequency ranges below 30 MHz.

III.4.11 Maximum Signal-to-Interference Ratio

The *maximum signal-to-interference ratio* defines the highest signal quality that can be reached at the AF output (Fig. III.31) with a reference output power suitable for an input signal modulated with nominal modulation (Section III.2). There is usually no accurate RF level quoted for the maximum SIR of a radio receiver, since in well-designed units the SIR remains the same for all signals received above a certain value. In poorly designed signal processing chains, however, it is possible that the SIR falls off when very high input levels are received.

When reaching the maximum SIR the AF harmonics are clearly higher than the noise signal, so that an evaluation with SINAD is very useful. Besides these linear distortions generated in the AF circuitry, other noise voltages occurring in the radio receiver or introduced via the power supply also have a limiting effect. The same is true for the close-in SSB noise (Section III.7.1) of the various heterodyning oscillators [6]. (In the

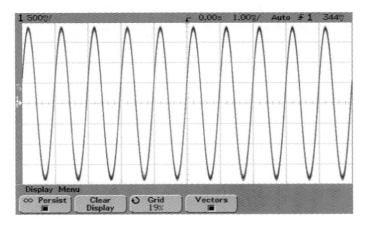

Figure III.31 Demodulated 1 kHz signal with maximum achievable SINAD at the AF output of a radio receiver. Compared with Figure III.22, the signal is clearly more stable and is subjected to hardly any changes. The fact that the signal has the same absolute amplitude as in Figure III.22 can be explained by the 'stabilizing' effect of the AGC in the receiver. (The black curves represent the output voltage measured at the time of observation. The light gray area is not much wider and shows the variations due to the influence of noise over an observation time of 10 s.)

evaluation of logical states during frequency keying or phase keying, delay differences are relevant as well.)

Some specifications prefer to evaluate the (demodulation) distortion factor (Section III.13.3) for one or several defined RF levels with nominal modulation at the reference output power instead of the maximum signal-to-interference ratio (Fig. III.32).

III.4.12 Measuring the Operational Sensitivity and Maximum SIR

The measuring setup for determining the operational sensitivity and the maximum signal-to-noise ratio is shown in Fig. III.33.

Table III.3 Subjective auditory impression with different SNR values according to [35]

Signal-to-noise ratio	Audibility
0 dB	Lower limit of perceptibility
10 dB	Lower limit of speech intelligibility
20 dB	Useful voice communication
30 dB	Sufficient music reproduction quality
40 dB	Broadcasting quality (mono)
50 dB	Useful broadcasting quality (stereo)
54 dB	Minimum requirements for hifi quality
60 dB	Very good music reproduction

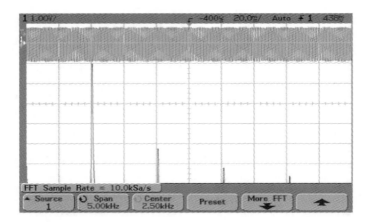

Figure III.32 Demodulated 1 kHz useful signal in the frequency range of a radio receiver with almost distortion-free modulation. The second harmonic is at a distance more than −42 dB from the useful signal. (Y axis: 10 dB/div., X axis: 500 Hz/div.; the upper part shows the demodulated signal in the time domain, but with another time base than that in Figure III.31).

Measuring procedure:

1. Tune the receiver to the frequency range to be tested.
2. Connect the SINAD meter (via a weighting filter if necessary) to the AF output of the receiver.
3. Tune the test transmitter to the receive frequency and feed the lowest possible RF level P_1 modulated with nominal modulation (Section III.2) to the receiver.
4. Increase the RF level P_1 until the desired SINAD (Section III.4.8) is obtained at the AF output. (Usually a SINAD of 10 dB, 12 dB or 20 dB is used as a quality criterion.)
5. Note the P_1 level.

The value P_1 obtained represents the operational sensitivity of the receiver under the measuring conditions used.

To obtain the maximum SIR the RF level P_1 is varied (not above 0 dBm) until the highest SINAD is reached.

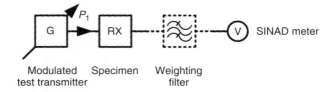

Figure III.33 Measuring arrangement for determining the operational sensitivity and the maximum SIR.

III.4.13 Measuring Problems

When determining the noise figure of the receiver by calculation using the measured minimum discernible signal, the noise components resulting from spurious signal reception (Section III.5) are of no importance. In practice, this is especially important with a specimen of insufficiently suppressed image frequencies or poor IF rejection (IF breakdown).

For measuring the minimum discernible signal and the operational sensitivity, the test transmitter must be effectively shielded and choked. Otherwise, radiation or falsification due to unwanted energy can enter the receiver via the housing surface or the power supply network. This should be checked by disconnecting the test transmitter from the antenna socket with a high test transmitter output level.

In cases in which the AF frequency response has been designed to decrease with higher frequencies, the measurement of the specimen provides results which are too good. (The acoustic reproduction sounds relatively dull). Measuring the noise figure then provides an objective criterion for evaluating the sensitivity.

It is frequently observed that, when switching to another IF bandwidth (if the unit offers this possibility), the SNR does not increase and the minimum discernible signal limit does not improve proportionately to the change in the bandwidth. If the AF bandwidth is dimensioned more generously than the IF bandwidth, there are various reasons for this. In receivers of modern design, selection takes place before the signal reaches the main amplifier. With narrow IF bandwidths, the broadband noise of the IF stages can have a significant influence [6]. Other causes may be the different (mis-)match of the various IF filters in use or their different passband attenuation.

For determining the maximum SIR by measurement, the spectral purity of the modulation signal is of great importance. Otherwise, the measurements on radio receivers with excellent demodulation properties can falsely change to the *detriment* of the specimen.

III.5 Spurious Reception

III.5.1 Origin of Inherent Spurious Response

Owing to insufficient screening of the assemblies in the receiver, clock signals, edge structures of digital signal pulses, mixing products of the various oscillators and their non-harmonics (Section III.7.2) can generate interfering signals having frequencies that are within the receive frequency range or identical to one of the intermediate frequencies. The propagation of such interferences through the supply voltages in the unit and crosstalk between signal lines can also be a source of this type of interference. These are called *inherent spurious response*, since they originate in the radio receiver.

When receiving class A1A and J3E emissions these become audible directly as 'whistling' or 'chirping' sounds. With emission class F3E, the noise level can be reduced relative to input frequencies without disturbances, while with class A3E emissions one may hear a bright sound with interjected chirping as the noise level increases. In all cases, inherent spurious response always reduces the SIR and sometimes even masks weaker signals. In this regard, commercially available radio receivers or receiver modules vary widely.

III.5.2   Measuring Inherent Spurious Response

The test setup for measuring inherent spurious response and determining its signal strength is illustrated in Figure III.34.

Measuring procedure:

1. Connect a dummy antenna to the antenna socket to prevent the introduction of external interferences.
2. Connect a voltmeter or an AF level meter to the AF output of the receiver.
3. Tune the receiver across the entire receive frequency range in increments adapted to the receive bandwidth (Section III.6.1) or according to the frequency pattern assigned.
4. Note the receive frequencies at which interferences are detected and the resulting AF output levels.
5. Tune the receiver very close to the inherent spurious response detected, so that this is no longer detectable.
6. Tune the test transmitter to the receive frequency, use nominal modulation (Section III.2), and increase the RF level P_1 until the AF output level noted in step 4 is achieved (substitution method).
7. Note the P_1 level in addition to the notes made in step 4.

P_1 represents the level of an RF input signal equivalent to the inherent spurious response detected. The strongest inherent spurious response detected is used for defining the equipment specification. The correct information in data sheets could read:

   'No inherent spurious response appears to be stronger than an equivalent input signal of $-110\,\mathrm{dBm}$.'

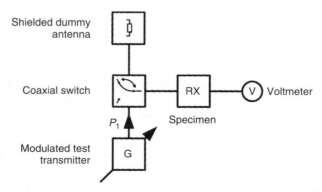

Figure III.34   Measuring arrangement for recording and determining the signal strength of inherent spurious response.

(Some manufacturer data relates to a useful signal with nominal modulation and a certain RF level that does not drop below a certain SINAD (Section III.4.8) due to an inherent spurious response at any receive frequency.)

The strength of the inherent spurious response can also be determined with an internal instrument for indicating the relative receive signal strength (Section III.18), provided that this instrument is sufficiently accurate.

III.5.3 Reception and Suppression of Image Frequencies

Converting a signal by mixing, as is done in every superhet receiver (Section I.2.2), produces no unambiguous results (Section V.4.2), because the following equations apply for a given intermediate frequency

$$f_{IF} = f_{LO} - f_{RX} \quad \text{with } f_{LO} \text{ higher than } f_{RX} \qquad (III.15)$$

$$f_{IF} = f_{RX} - f_{LO} \quad \text{with } f_{LO} \text{ lower than } f_{RX} \qquad (III.16)$$

where
f_{IF} = (first) intermediate frequency, in Hz
f_{LO} = LO injection frequency, in Hz
f_{RX} = receive frequency, in Hz

since for the mixer it is basically irrelevant whether the frequency fed to the LO port is higher or lower than the frequency fed to the RF port. If a signal having *a frequency that differs from the receive frequency by double the IF* is present at the receiver input, then the criterion for conversion to the IF is met once again

$$f_{IF} = (f_{RX} + 2 \cdot f_{IF}) - f_{LO} = f_{image} - f_{LO} \quad \text{with } f_{LO} \text{ above } f_{RX} \qquad (III.17)$$

$$f_{IF} = f_{LO} - (f_{RX} - 2 \cdot f_{IF}) = f_{LO} - f_{image} \quad \text{with } f_{LO} \text{ below } f_{RX} \qquad (III.18)$$

where
f_{IF} = (first) intermediate frequency, in Hz
f_{RX} = receive frequency, in Hz
f_{LO} = LO injection frequency, in Hz
f_{image} = image frequency, in Hz

This is the *image frequency* (Fig. III.35), because 'for the mixer it is basically irrelevant whether the frequency fed to the LO port is higher or lower than the frequency fed to the RF port'.

If a station of, for example, 92.1 MHz tuned in is mixed in a VHF broadcast receiver to an IF of 10.7 MHz using a local oscillator of 102.8 MHz, then the image frequency is

$$f_{image} = 2 \cdot 10.7 \text{ MHz} + 92.1 \text{ MHz} = 113.5 \text{ MHz}$$

If a signal reaches the mixer at this frequency of 113.5 MHz, it will also appear at the intermediate frequency, since

$$f_{IF} = 113.5 \text{ MHz} - 102.8 \text{ MHz} = 10.7 \text{ MHz}$$

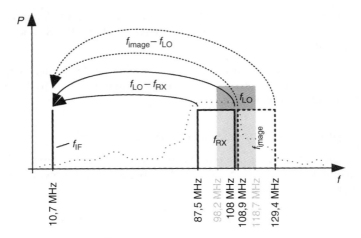

Figure III.35 Ambiguity of the heterodyning principle, illustrated with VHF FM broadcast frequencies. With the LO injection signal, which can be tuned between 98.2 MHz and 118.7 MHz, the entire VHF broadcast band is converted to a fixed IF of 10.7 MHz. Suppressing the image frequencies is difficult because of the proximity to the frequency band of the useful signal, since RF bandpass filters have a finite edge steepness (dotted line).

But if the VHF broadcast receiver has an LO injection signal of 81.4 MHz for mixing to the IF (f_{LO} is now below f_{RX}), image the frequency will be

$$f_{\text{image}} = 92.1\,\text{MHz} - 2 \cdot 10.7\,\text{MHz} = 70.7\,\text{MHz}$$

since a signal at this frequency will result again in

$$f_{\text{IF}} = 81.4\,\text{MHz} - 70.7\,\text{MHz} = 10.7\,\text{MHz}$$

and thus assumes the intermediate frequency. To prevent interference, it is therefore necessary to sufficiently suppress image frequency ranges by filter(banks) before the mixer. With a receive frequency range wider than $2 \cdot f_{\text{IF}}$ the receive frequency and the image frequency ranges will overlap [21]. Suppressing the image frequency without affecting the entire desired frequency range, as required, makes it necessary to design the RF selection as a switchable input bandpass filter or a tunable bandpass filter (RF preselector) of

$$Q > \frac{f_{\text{RX}}}{f_{\text{IF}}} \tag{III.19}$$

where
  Q = operational quality (Section III.11) of the RF selection tuned to the receive
         frequency, dimensionless
  f_{RX} = receive frequency, in Hz
  f_{IF} = (first) intermediate frequency, in Hz

The greater the intermediate frequency, the further apart are the image frequency and the receive frequency, making the filter circuit either less complex or, for the same level of complexity, more efficient (Fig. III.36).

For radio receivers that cover a wide receive frequency range it is therefore important to have a high IF. Nowadays, IF blocks can be realized that use several hundred MHz or even GHz, as found in broadband measuring/test receivers. Owing to the narrow IF filter bandwidths required for many emission classes but not sufficiently achievable, mixing down to an additional lower IF is necessary. The actual near selection (Section III.6) for narrow IF bandwidths must be performed on the second IF by high-quality filters. This causes image frequency reception on the second IF. Its suppression must be ensured by the attenuation effect of the selector used in the first high IF stage.

III.5.4   IF Interference and IF Interference Ratio

Owing to the limited insulation between the RF port and IF port of mixers (Fig. III.37) and the crosstalk within the individual receiver modules, the signals can reach the IF level(s) *directly* without being converted. This phenomenon is called *IF interference* ('disruptive IF breakdown') [21]. Other terms found in the relevant literature are IF immunity and IF rejection.

If the IF interference is insufficiently attenuated, such interferences are audible across the entire receive frequency range and are therefore particularly troublesome. A change in the

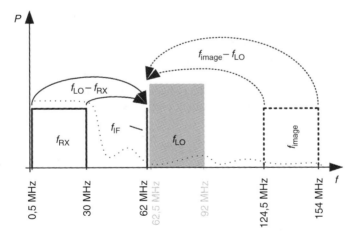

Figure III.36   Image frequency position with high (first) IF, illustrated with the frequency plan of an MF/HF receiver. With the tunable LO injection signal, the entire receive spectrum between 500 kHz and 30 MHz is converted to the high intermediate frequency of 62 MHz, which is higher than the receive frequency. Even a low-pass filter of moderate order having a limit frequency above the highest receive frequency is sufficient to adequately suppress image frequencies (dotted line). In addition, the frequencies of the LO injection signal are effectively attenuated in order to minimize interfering stray radiation (Section III.17) emitted by the receiver.

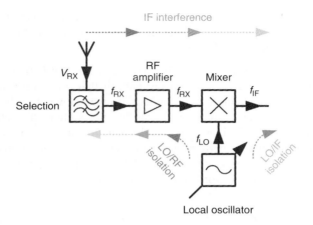

Figure III.37   The ports of a real mixer are not fully isolated from each other. In addition there is some crosstalk between the receiver modules, so that unwanted signals from/to the antenna socket occur in/from the signal path.

receive frequency brings no improvement. If a signal of a frequency equal to the lowest IF used in the receive path is present at the antenna socket [22], a receiver with insufficient IF rejection can process this signal like a tuned radio frequency receiver (Fig. I.2).

III.5.5   Reception of Other Interfering Signals

With sufficiently high input levels harmonics of the input signal arise in the mixer and have an interfering effect as well. In such cases the harmonics of the input signal may be converted to the IF together with the fundamental wave and the harmonic waves of the LO injection signal [21]. In practical applications it may be advantageous for detecting such spurious receiving frequencies to replace the LO frequency by the IF frequency position (e.g., taken from the receiver data sheet) in Equation (V.26) and to derive the spurious frequency from the receive frequency selected by

$$f'_{RX} = \left| \frac{n}{m} \cdot f_{RX} + \frac{n \pm 1}{m} \cdot f_{IF} \right| \tag{III.20}$$

where
f'_{RX} = spurious receive frequency caused by harmonic mixing, in Hz
$\quad n = 0, 1, 2, 3, \ldots$
$\quad m = 1, 2, 3, \ldots$
f_{RX} = receive frequency, in Hz
f_{IF} = (first) intermediate frequency, in Hz

Particularly critical is mixing the second harmonic of the receive signal with the second harmonic of the LO injection signal. This interference is spaced to the selected receive frequency by half the intermediate frequency position. Especially this type of interference exists in wideband RF receivers that have a high IF position but no input bandpass filter (only a low-pass filter for the image frequency). This is the reason why VHF/UHF receiving systems having a low first IF can be designed with an input bandwidth of only half the IF frequency position, that is, a maximum of only 5 MHz at an IF of 10.7 MHz.

Based on the development process, the relation between these interferences and their interaction can be mathematically determined. This will be rigorously explained with mathematical formulas in Section V.4, while it is described here verbally for easy comprehension.

Often, all these effects described in Sections III.5.3 and III.5.5 are collectively called *spurious signal reception*.

III.5.6 Measuring the Spurious Signal Reception

The measuring setup for determining the spurious signal reception is illustrated in Fig. III.38.

Measuring procedure:

1. Tune the receiver to the frequency range to be tested.
2. Connect the SINAD meter to the AF output of the receiver.
3. Modulate the test transmitter with nominal modulation (Section III.2) and supply an RF level P_1 of 100 dB above the operational selectivity for 12 dB SINAD (Section III.4.8) to the receiver.
4. Tune the test transmitter (in increments adapted to the used receive bandwidth) from the lowest IF to the highest relevant frequencies. (To shorten the procedure it is possible to use the intermediate frequencies and the first image frequency of the specimen only.)
5. With frequencies that show an increase in the AF output signal, decrease P_1 until a SINAD of 12 dB is achieved.
6. Note the respective frequency of the test transmitter and the P_1 level.

The difference between P_1 and the operational sensitivity at 12 dB SINAD represents the suppression of the relevant spurious signal reception.

III.5.7 The Special Case of Linear Crosstalk

Depending on the circuit design of a J3E receiving path, the undesired sideband can appear more or less prominent in the demodulated channel. If a strong RF input signal falls into

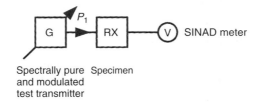

Spectrally pure   Specimen
and modulated
test transmitter

Figure III.38 Measuring arrangement for determining the spurious signal reception.

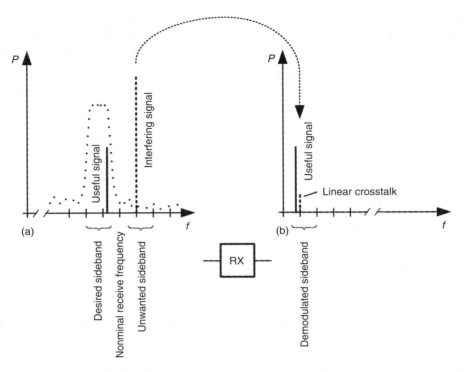

Figure III.39 If a strong interfering signal together with a useful signal appears on the receiver input in the frequency range of the unwanted sideband, as shown in graph (a), then the interfering signal can be noticed as linear crosstalk in the demodulated AF signal as shown in graph (b). (The dotted line in (a) schematically depicts the selection curve of the IF sideband filter.)

the receive frequency range of the suppressed sideband (for example 20.001 MHz) (it becomes mirrored to the desired sideband, whereby the nominal receive frequency is the centre frequency), it can affect the reception in the demodulated sideband (19.999 MHz in this example) (Fig. III.39).

This can be regarded as another type of spurious signal reception. This effect is known as linear crosstalk [23]. In many cases this is caused by the fact, that the demodulator used has a finite balance. In present receiver concepts, linear crosstalk is no longer very important as the suppressive action is so strong, that other interfering events affect the reception earlier or are more prominent.

III.5.8   Measuring the Linear Crosstalk Suppression

The measuring setup for determining the suppression of the linear crosstalk is shown in Figure III.40. It is normal practice to perform this measurement separately for the class J3E emission as indicated below.

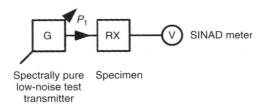

Figure III.40   Measuring arrangement for determining linear crosstalk.

Measuring procedure:

1. Tune the radio receiver to the frequency range to be tested.
2. Connect the SINAD meter to the AF output of the receiver.
3. Tune the test transmitter to a frequency offset relative to the receive frequency for which a 1 kHz tone is expected after demodulation.
4. Increase the output level P_1 of the test transmitter until a SINAD of 12 dB is obtained.
5. Note the P_1 level (this corresponds to the operational selectivity for 12 dB SINAD).
6. Tune the test transmitter in the opposite direction by the amount of the frequency offset in step 3 (that is, to the suppressed sideband).
7. Increase the output level P_1 of the test transmitter until a SINAD of 12 dB is obtained again.
8. Note the P_1 level.

The difference between the two recorded P_1 levels corresponds to the suppression ratio of the linear crosstalk.

III.5.9   Measuring Problems

When determining the spurious signal reception by measurement the spectral purity of the test transmitter used is of great importance, since the spurious waves of the measuring signal supplied can *falsely simulate* spurious signal reception. The method using a modulated measuring signal as described in Section III.5.6 can be helpful, if only to a certain degree.

For the determination of spurious signal reception with very high suppression (better than 90 dB) it must be considered that the measuring signal can drive the frontend into compression (Section III.8.1). This will falsify the readings to the *advantage* of the specimen.

When determining the linear crosstalk, the spectral purity of the test transmitter used is of great importance. Often, reciprocal mixing (Section III.7) will occur much earlier than the adverse effect of the linear crosstalk. In receiving paths reasonably in keeping with the state of the art, for J3E the SSB noise can clearly exceed the interfering tone resulting from linear crosstalk. In such cases, there is in fact no reasonable way to measure the linear crosstalk. This means that a SINAD of 12 dB, as visualized in Section III.5.8 at step 7, cannot be obtained.

III.6   Near Selectivity

Near selectivity serves for separating the receive signal from unwanted signals and noise outside the receive channel. It is mainly determined by selectors used at the IF level (for example, electromechanical filters (Fig. III.41), quartz filters, DSP filtering by digitally simulated stages). In addition, AF filters can be used to minimize the noise components resulting from demodulation. They also favour certain frequency ranges in the AF signal mixture (Section III.13). Such AF filters can be built easily and cheaply and can be connected to the receiver externally.

In contrast, narrow IF filters built into the upstream IF stages will relieve signal processing stages downstream. This has a positive effect on several parameters relating to the receiving characteristics. The suppression of adjacent channels described in this section is mainly performed by the stop-band attenuation of the IF filter employed.

The time differences occurring in narrow-band frequency spectra (frequency groups) in the passband range of the selector are called the filter group delay. Figure III.43 shows group delays typical for electromechanical filters. DSP filtering by digitally simulated stages is mostly performed with finite impulse response (FIR) filters of symmetrical coefficients for near selection. Their filter group delay is constant, so that there is no group delay distortion. The delay in the frequency response is

$$t_{gr} = \frac{N - 1}{2 \cdot f_{s\,FIR}} \qquad (III.21)$$

where

t_{gr} = group delay time of a FIR filter with symmetrical coefficients, in s
N = number of taps, dimensionless
$f_{s\,FIR}$ = sampling rate at the FIR filter, in S/s

When demodulating digital (especially phase-modulated) emission classes, the group delay time differences have a limiting effect on the achievable bit error rate (Section III.4). The

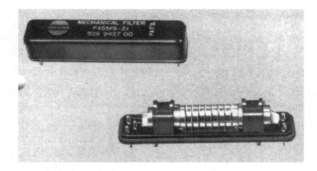

Figure III.41   Electromechanical filter as used for near selection in the second or third IF of high-quality VLF/HF radio receivers with analog signal processing. (Company photograph of Rockwell-Collins.)

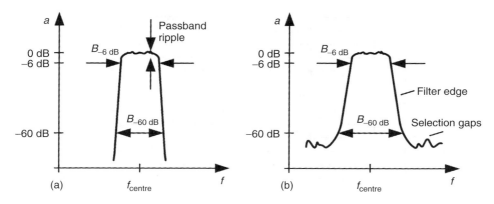

Figure III.42 Important characteristic parameters regarding the selectivity are the bandwidth $B_{-6\,dB}$ of the receiving path, ripple in the passband region, and the attenuation figure in the cutoff region. Depending on the type of filter, there can be so-called selection gaps due to insufficient far-off selection properties of the filter. Graph (a) shows the attenuation curve for a low (good) shape factor, and graph (b) shows the curve for a poorer shape factor. While the $-6\,dB$ bandwidth is the same, the $-60\,dB$ bandwidth differs greatly.

filter group delay is measured with filters incorporated in the test circuit with vectorial network analyzers.

III.6.1 Receive Bandwidth and Shape Factor

The *receive bandwidth* is determined mainly by the emission class to be demodulated and is expressed as the frequency range for which the passband attenuation figure has increased by 6 dB (sometimes 3 dB only). Within this $-6\,dB$ bandwidth is the passband range, and above this is the cutoff region. The receive bandwidth (Section III.4.4) is always an important parameter in relation to all characteristic values based on the receiver sensitivity (Section III.4).

The edge steepness (attenuation increase in the cutoff region) also influences the *shape factor*. This is the ratio between the $-60\,dB$ bandwidth and $-6\,dB$ bandwidth, sometimes also called the form factor. The closer its value approaches 1, the more rectangular is the selection curve. Signals just outside the filter edges are suppressed completely (Fig. III.42). High-quality radio receivers allow the determination of bandwidths of up to $-80\,dB$. It is then reasonable to quote the shape factor calculated from $-80\,dB$ bandwidth to $-6\,dB$ bandwidth as an indication of the radio receiver performance. Especially with applications in which no exactly defined frequencies are assigned to the transmission channel, the shape factor is an important quality parameter and, for specification purposes, is preferred over the adjacent channel suppression.

Very high edge steepness and very low shape factors close to 1 (almost rectangular selection curves with a sharp-edged transition from the passband region to the cutoff region) produce an unnatural sound pattern. DSP filtering with FIR filter in digitally simulated stages allows the control of the slope steepness and the shape factor by selecting

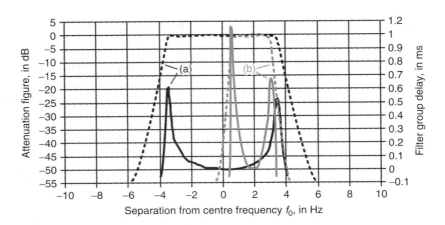

Figure III.43   Graph (a) shows the filter group delay of an electromechanical IF filter with 7 kHz bandwidth for A3E reception, and graph (b) shows the filter group delay of a 2.5 kHz filter for J3E reception. The dotted lines represent the corresponding selectivity curves. There is a marked rise of the delay at the transition from the passband region to the cutoff region. Narrow filter bandwidths and low shape factors cause bigger differences in the group delays.

different window functions for implementing the filter (windowing). (The window function determines the weighting of the sampling value inside a segment, the window, for the calculation. With a window function the signal segment intended for the fast Fourier transformation (FFT) can be more or less eliminated or attenuated towards zero. This improves the attenuation in the cutoff region [24].) The sound pattern obtained with the Blackman-Harris window, which offers excellent attenuation in the cutoff region with a (for digital filters) moderately reduced shape factor (of e.g., ~1.3), results in a more natural sound impression than a Kaiser window, with its steeper filter edges.

For good speech reproduction and natural sound, the product of the two −6 dB selection points should be approximately 500,000. Using a 2.7 kHz filter with a passband region between 175 Hz and 2.87 kHz for J3E reception would meet this requirement, since $175 \cdot 2{,}870 = 502{,}250$. (If a sufficiently broad bandwidth is available from, for example, coordinated frequency assignments with large enough channel spacing, the value can be extended up to 1,200,000. In such instances, the use of a −6 dB selection point having a frequency that is not too low (~300 Hz) is recommended. This would satisfy the requirement of a frequency response for A3E voice communication used by most air traffic control institutions with a passband region of 300 Hz to 3.4 kHz and the resulting product of $300 \cdot 3{,}400 = 1{,}020{,}000$.)

III.6.2   Measuring the Receive Bandwidth

The measuring setup for determining the receive bandwidth is shown in Figure III.44.

Measuring procedure:

1. Tune the receiver to the frequency range to be tested.
2. Tune the test transmitter to the frequency at the centre of the IF passband region.

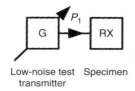

Figure III.44 Measuring arrangement for determining the receive bandwidth.

3. Adjust the RF level P_1 to obtain S5 (Section III.18).
4. Increase P_1 by 6 dB.
5. Adjust the test transmitter frequency by decreasing ($f_{low\,6\,dB}$) and increasing ($f_{up\,6\,dB}$) so that each time S5 is obtained exactly again. Note the frequencies.
6. Increase P_1 by 60 dB above the setting in step 3.
7. Change the test transmitter frequency by decreasing ($f_{low\,60\,dB}$) and increasing ($f_{up\,60\,dB}$) until S5 is obtained exactly again. Note the frequencies.

For radio receivers having no indicator for the receive signal strength, the AF output level can be used similarly to that for class A1A and J3E emissions to determine the −6 dB bandwidth. In this case it is essential to deactivate the AGC. The *receive bandwidth* is the difference between the two frequencies $f_{up\,6\,dB}$ and $f_{low\,6\,dB}$ determined. The shape factor is derived from the ratio of the differences between the frequencies determined for the respective attenuation figure with

$$SF = \frac{f_{up\,60\,dB} - f_{low\,60\,dB}}{f_{up\,6\,dB} - f_{low\,6\,dB}} = \frac{B_{-60\,dB}}{B_{-6\,dB}} \tag{III.22}$$

where
SF = shape factor describing the near selectivity of the receiving path, dimensionless
$f_{up\,60\,dB}$ = upper −60 dB selection point determined for the receiving path, in Hz
$f_{low\,60\,dB}$ = lower −60 dB selection point determined for the receiving path, in Hz
$f_{up\,6\,dB}$ = upper −6 dB selection point determined for the receiving path, in Hz
$f_{low\,6\,dB}$ = lower −6 dB selection point determined for the receiving path, in Hz
$B_{-60\,dB}$ = −60 dB bandwidth of the receiving path, in Hz
$B_{-6\,dB}$ = receive bandwidth (−6 dB bandwidth) of the receiving path, in Hz

Example: If the radio receiver is tuned to 6.050 MHz for class A3E emission and an S meter indication of S5 is obtained after the 6 dB increase of the test transmitter level P_1 at a frequency of 6.048,71 MHz and 6.051,5 MHz, the receive bandwidth is then 2.79 kHz. Furthermore, if after the increase of P_1 by another 54 dB the S5 reading is obtained at the frequencies 6.047,81 MHz and 6.052,53 MHz, then an entirely acceptable shape factor is

$$SF = \frac{6.052,53 \text{ kHz} - 6.047,81 \text{ kHz}}{6.051,5 \text{ kHz} - 6.048,71 \text{ kHz}} = 1.7$$

This suggests an acceptable electromechanical selection. Taking a closer look at the spacing of the frequency positions at the selection points from the centre frequency, it is then apparent that the offsets below and above are slightly different. This indicates less than optimal adjustment (asymmetry) of the IF filter centre frequency.

III.6.3   Adjacent Channel Suppression

Adjacent channel suppression describes a certain decrease in the useful signal-to-interference ratio of a demodulated received signal to an (interfering) signal in the adjacent channels assigned above and below the receive channel. It defines the ratio between the useful signal level increased by the nominal modulation (Section III.2) and the level of the signal causing the SIR decrease in the adjacent channel. The term *dynamic selectivity* of radio receivers is sometimes used and in some literature the name two-signal selectivity and adjacent channel selectivity or adjacent channel ratio can be found. Adjacent channel suppression is used as a quality parameter, especially in applications and radio services in which exact frequencies and channel pattern are assigned to the transmission channels used.

A similar convention can be applied for class J3E emissions, for which an unmodulated (interfering) signal is measured with a signal spacing of $\pm 3\,\text{kHz}$ above and below the used carrier frequency.

III.6.4   Measuring the Adjacent Channel Suppression

The measuring setup for determining the adjacent channel suppression is shown in Figure III.45.

Measuring procedure:

1. Tune the receiver to the frequency range to be tested.
2. Connect the SINAD meter to the AF output of the receiver.
3. Tune the test transmitter, and with it the receiver, to the receive frequency, while observing the attenuation figure of the power combiner. Supply P_{use} with nominal

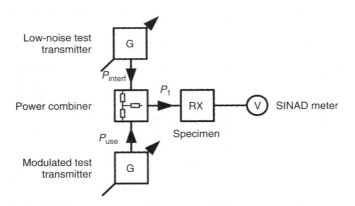

Figure III.45   Measuring arrangement for determining the adjacent channel suppression.

modulation (Section III.2) of an RF level resulting in a SINAD (Section III.4.8) of 12 dB at the AF output.

4. Increase the RF level P_{interf} of a second test transmitter within the frequency spacing of the assigned channel pattern below and above the receive frequency until the SINAD value has decreased by or to 6 dB.

The difference between P_{interf} and P_{use} corresponds to the adjacent channel suppression, relative to the respective neighbouring channel below or above the receive channel.

III.6.5 Measuring Problems

A high SSB noise (Section III.7.1) of the LO injection signal close to the carrier frequency renders the determination of the −60 dB selection point impossible, even with sufficient filter attenuation. At best, a usually instable indication of the receive signal strength (Section III.18.1) can be detected during the measuring procedure. As the graph in Figure III.46 shows, the carrier peak at the filter edge is suppressed by the attenuation figure. The noise components within the filter passband remain unattenuated (shaded area in the graph). This suggests an inadequate design of the oscillator and produces false results for the selection points determined.

Important for selectivity tests is the noise characteristic of the test transmitter producing the signal P_{interf} in Figure III.45 [25]. If the SSB noise ratio (Section III.7.1) is insufficient, the measurement does not determine the actual selection but an erroneous value *to the disadvantage* of the specimen. Hardly more than adequate accuracy definable adjacent channel suppression of a test transmitter of known single sideband noise can be determined by:

$$a_{\text{adj}}(_{\Delta}f) = -\left(L_{\text{TTX}}\left(_{\Delta}f\right) + SINAD + 2\,\text{dB} + 10 \cdot \lg\left(\frac{B_{-6\,\text{dB}}}{1\,\text{Hz}}\right)\right) \qquad (\text{III}.23)$$

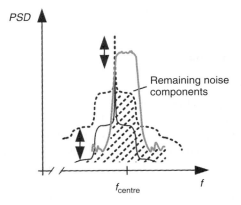

Figure III.46 Signal patterns at the IF level during the determination of the shape factor in a receiving path with high SSB noise of the LO injection signal close to the carrier frequency. Signals outside the filter passband region are suppressed by the appropriate stopband attenuation figure. Signal components (noise in this example) present in the filter passband region together with the signal to be measured remain at full strength.

where

$a_{adj}(_{\Delta}f)$ = maximum definable adjacent channel suppression, depending on the frequency spacing $(_{\Delta}f)$, in dB

$L_{TTX}(_{\Delta}f)$ = SSB noise ratio of the test transmitter used, depending on the carrier frequency spacing $(_{\Delta}f)$, in dBc/Hz

SINAD = SINAD across which the adjacent channel suppression of the radio receiver is determined, in dB

$B_{-6\,dB}$ = receive bandwidth (-6 dB bandwidth) of the receiving path, in Hz

The introduction of a safety factor of 2 dB provides sufficient reserves, even for receiving paths with an inadequate shape factor.

If, for example, a test transmitter with -135 dBc/Hz at 25 kHz SSB noise ratio is available, then it allows at best the determination of an adjacent channel suppression of

$$a_{adj}(_{\Delta}f = 25\ \text{kHz}) = -\left(-135\ \text{dBc/Hz} + 12\ \text{dB} + 2\ \text{dB} + 10 \cdot \lg\left(\frac{15,000\,\text{Hz}}{1\,\text{Hz}}\right)\right)$$

$$= 79\ \text{dB}$$

in an F3E receiving path designed for a 25 kHz channel pattern when following the measuring method described in Section III.6.4.

In some cases the SINAD decreases almost continuously with an increasing interference signal in the adjacent channel (P_{interf} in Fig. III.45) up to a certain level. If the interfering signal increases further, the SINAD then begins to improve (rises) again. This is repeatedly experienced in F3E receiving paths. It can be explained in the sense of a basic blocking effect (Section III.8). Before this happens, however, the SINAD almost always decreases by or to 6 dB, as is desired for determining the adjacent channel suppression. During the measuring procedure, it is important to actually vary the interfering signal from smaller to higher levels and to observe the changes in the reading of the SINAD meter. This prevents any erroneous interpretations from the start.

III.7   Reciprocal Mixing

III.7.1   Single Sideband Noise

If the output signal of an ideal oscillator were to be displayed in the frequency range (e.g., by a spectrum analyzer) the result would be a single narrow line having a height corresponding to the output amplitude, that is, to the output power at a defined resistance. This is however not the case under real conditions, since additional power components are measured with varying separation from the oscillator frequency. They become weaker with an increasing frequency separation (Fig. III.47). The reasons lie in the finite frequency and amplitude stability of the oscillator, so that it becomes modulated by its own instability. The measurement of these power components, called *noise sidebands*, essentially describes the short-time stability of the oscillator.

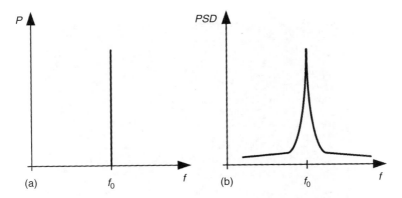

Figure III.47 Graph (a) shows the output spectrum of an ideal oscillator, and graph (b) shows that of a real oscillator.

The noise sidebands decrease equally on both sides of the carrier. The description of their performance quality is limited to the power of one noise sideband at a frequency spacing. The term *single sideband noise (SSB noise)* or simply oscillator noise is commonly used. To enable an objective comparison of specifications it is reasonable to relate these to a measuring bandwidth of one Hertz (dBc/Hz). This expresses a spectral noise power density (Section III.4.1). See also the facts outlined in Section III.4 and particularly in Section III.4.4. A technically correct specification would be

$- 95\,\mathrm{dBc/Hz} @ 10\,\mathrm{kHz}$

or

$- 95\,\mathrm{dBc/Hz}$ with $10\,\mathrm{kHz}$ separation

and means that the power of the single sideband noise at a frequency spacing of 10 kHz to the carrier is 95 dB below the carrier peak in a measuring bandwidth of 1 Hz. The procedure for performing such measurements directly at the signal source is described in [8] and in current application brochures from several manufacturers of high-end measuring equipment (Fig. III.48).

The actual progression, or the type of SSB noise decrease with an increasing frequency is dominated by the type of frequency preparation (e.g., with LC oscillators – free-ranging or coupled to a PLL, direct or indirect synthesizer, quartz oscillator). An optimum can be achieved by an adequate circuit design, while the total elimination of the SSB noise cannot be 'enforced'. Temperature-controlled (heated) quartz oscillators (TCXO or OCXO) have excellent properties. Their output frequency allows in return no or only very little variation, which makes these oscillators useful in modern receiver engineering only as injection oscillators in the second IF stage or as reference oscillators for a PLL (Fig. I.8) or DDS

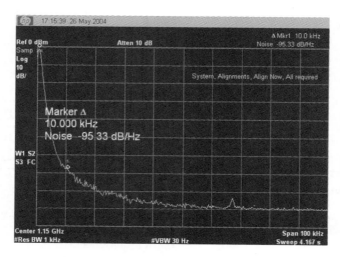

Figure III.48 Single sideband noise (or SSB noise ratio) at the signal source determined by a modern spectrum analyzer almost at the push of a button and without any conversions. The performance of the spectrum analyzer must, however, be significantly better than that of the specimen. The display shows an arbitrary sinusoidal signal at 1.15 GHz. Only the upper noise sideband up to 100 kHz separation from the carrier frequency is shown.

(Section I.2.2). If the signal is obtained by multiplication of a base frequency, the SSB noise ratio decreases to

$$L_{mult}(_\Delta f) = L_{f0}(_\Delta f) + 20 \cdot \lg(n) \tag{III.24}$$

where
$L_{mult}(_\Delta f)$ = SSB noise ratio after multiplication, depending on the carrier frequency
    separation ($_\Delta f$), in dBc/Hz
$L_{f0}(_\Delta f)$ = SSB noise ratio of the fundamental signal used for multiplication,
    depending on the carrier frequency separation ($_\Delta f$), in dBc/Hz
n = multiplication factor, dimensionless

A rule of thumb for practical work is that with every frequency doubling by multiplication the SSB noise ratio decreases by 6 dB. Conversely, the value increases with frequency division. When using the above-mentioned signal with an SSB noise ratio of −95 dBc/Hz as an example, then the noise ratio decreases according to Equation (III.24) by doubling the frequency (that is, with multiplication by factor 2) to

$$L_{mult}(_\Delta f = 10 \, kHz) = -95 \, dBc/Hz + 20 \cdot \lg(2) = -89 \, dBc/Hz$$

with the same frequency separation from the carrier.

As outlined above, the SSB noise is caused by the amplitude and short-time frequency instability, the latter of which is related to the phase instability (Fig. III.49). The SSB noise

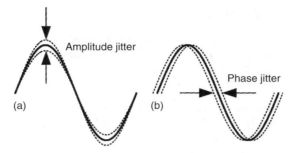

Figure III.49   Graph (a) shows the amplitude noise and graph (b) the phase noise within the time domain.

is therefore comprised of these two components: In the region close to the carrier (often less than a few 100 Hz apart, but sometimes up to several ten kHz [5]), the phase noise can dominate. The neighbouring noise is often caused by amplitude noise (Fig. III.50). Depending on the quality of the components determining the frequency and the limiting mechanisms in the circuit design of the frequency-generating oscillator stage, their spectral power density will vary to a higher or lower degree.

Both components can be measured separately by special measuring systems which, due to their high initial cost, are available in only a few laboratories. For this reason, specification parameters almost always state the *sum* of the noise power components. It should be noted that the term *phase noise* or single sideband *phase* noise should be used with caution for the reasons stated.

In cases where the oscillator noise is strongly influenced by the components of the frequency instability or phase jitter, the interfering deviation (or residual FM) is also used instead of the SSB noise ratio. A direct conversion is possible by integration and is explained in [21].

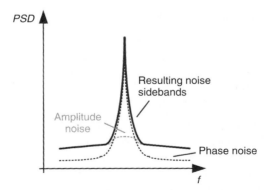

Figure III.50   The noise sideband resulting from the superimposed amplitude noise and phase noise.

III.7.2   Non-Harmonic (Close to Carrier) Distortions

Generated signals can be accompanied by interfering frequencies close to the carrier which project above the noise sidebands (Fig. III.51). These are called *spurious signals*. Their frequencies are not defined, since they appear above and below the carrier with differing frequency offsets.

A non-harmonic spurious signal can be thought of as a superposition of amplitude modulation and frequency modulation. This means that an FM of the carrier takes place with a modulation frequency equal to the frequency spacing between the carrier and the spurious component [25]. This has a direct effect on the short-time stability of the signal. Effects of this type can be particularly annoying when the signal source is used, for example, as the local oscillator in a receiver.

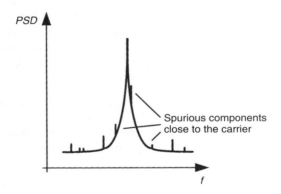

Figure III.51   Output signal of a real signal source, showing strong spurious components.

III.7.3   Sensitivity Reduction by Reciprocal Mixing

If various neighbouring signals are supplied to the receiver input, these pass the preselection and, after a possible RF amplification, are converted in the (first) mixer stage to the intermediate frequency by mixing (Section V.4) with the LO injection signal. During the mixing process the noise sidebands of the LO injection signal are transferred to the converting (receive) signals (Fig. III.52).

If a strong interference signal exists close to the receive signal (as shown in Fig. III.52), then the channel used becomes desensitized by the sideband noise transferred. The noise level in the receive channel increases, while the SIR decreases! In extreme cases, weaker signals may even be masked by the 'noise jacket'. The noise components falling within the receive bandwidth (Section III.6.1) increase the noise figure (Section III.4.2) of the receiver and can be quantitatively described (Fig. III.53). The actual reduction in sensitivity depends on the amplitude and the frequency spacing of the signals and, of course, on the SSB noise of the LO injection signal.

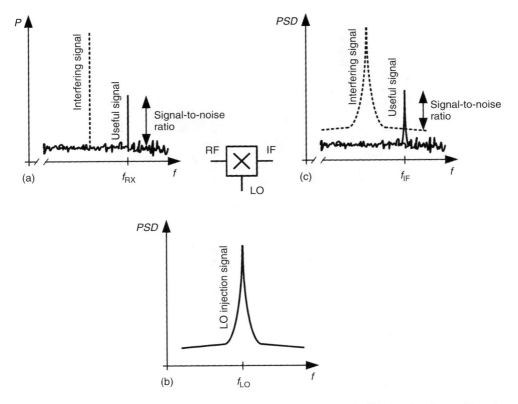

Figure III.52 When converting the input frequency spectrum to the IF, the signals reaching the mixer take over the noise sidebands of the LO signal. Graph (a) shows the signal scenario at the antenna socket, and graph (b) shows the LO injection signal supplied, with its noise sidebands. Graph (c) illustrates how the SSB noise transferred to the strong interference signal desensitizes the receive channel at the IF level.

The example contained in Figure III.53 assumes an interfering signal reaching the mixer with $-20\,\text{dBm}$ (vertical). With the LO injection signal having an SSB noise of $-120\,\text{dBc/Hz}$ and a frequency spacing between the useful signal and the interference signal, an additional noise figure of $F_{\text{dB RM}} = 34.5\,\text{dB}$ (intersection) arises. The actual resulting receiver noise figure, when considering the sensitivity reduction due to reciprocal mixing, is given by

$$F_{\text{dB res}} = 10 \cdot \lg\left(10^{\frac{F_{\text{dB}}}{10}} + 10^{\frac{F_{\text{dB RM}}}{10}}\right) \tag{III.25}$$

where

$F_{\text{dB res}}$ = resulting receiver noise figure when considering the sensitivity reduction due to reciprocal mixing, in dB

F_{dB} = receiver noise figure, in dB

$F_{\text{dB RM}}$ = additional noise figure caused by reciprocal mixing in Figure III.53, in dB

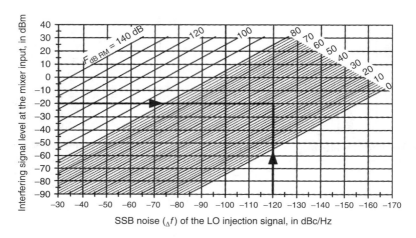

Figure III.53 The additional noise figure ($F_{dB\,RM}$) in the channel used is caused by reciprocal mixing and depends on the level of interference and the SSB noise of the LO injection signal for a certain frequency separation [6].

In this example, the receiver noise figure in a receiver with a specified noise figure $F_{dB} = 20\,dB$ is

$$F_{dB\,res} = 10 \cdot lg \left(10^{\frac{20\,dB}{10}} + 10^{\frac{34.5\,dB}{10}} \right) = 34.7\,dB$$

In this case, the total sensitivity is dominated almost exclusively by the noise component generated by reciprocal mixing in the receive channel.

The following rule of thumb applies to practical work: As long as $F_{dB\,RM}$ is at least 6 dB below the receiver noise figure, the sensitivity loss due to reciprocal mixing remains below 1 dB. If $F_{dB\,RM}$ and F_{dB} are equally high, the sensitivity drops by 3 dB. If $F_{dB\,RM}$ is at least 10 dB greater than F_{dB}, no calculation is necessary because the sensitivity of the receiver is influenced almost entirely by $F_{dB\,RM}$ (see sample calculation and Fig. III.54).

As with the noise sidebands, the spurious components of the LO injection signal are transferred by reciprocal mixing to the signal to be converted. During the A1A or J3E demodulation they become manifest as a whistling sound varying with the interference signal. In any case, when present in the useful channel they reduce the signal-to-interference ratio.

Of importance for the design of a multiple-conversion receiver (Section I.2.2) is primarily the LO injection signal of the first mixer stage, since up to this point in the receiving path no near selection takes place. Here, the effects described have their full impact. In the second mixer stage they can be largely avoided because of the earlier selection in the first IF stage with its relieving effect and because of the more easily achieved fixed-frequency generation for this LO injection signal.

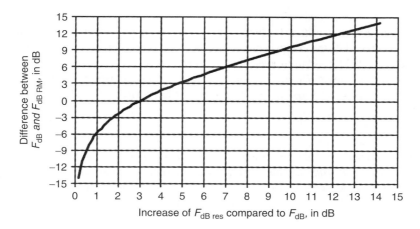

Figure III.54 Increase of the resulting receiver noise figure $F_{dB\,res}$ compared to the receiver noise figure F_{dB} by reciprocal mixing. The horizontal axis shows the difference between the receiver noise figure F_{dB} and the noise figure resulting from reciprocal mixing in Figure III.53. (In practice, a value of $-6\,dB$ can be regarded as, for example, an $F_{dB} = 20\,dB$ and an $F_{dB\,RM} = 14\,dB$.) The diagram is generally applicable and thus renders the calculations with Equation (III.25) unnecessary. If $F_{dB\,RM}$ exceeds the receiver noise figure F_{dB} by more than $10\,dB$ (upper right corner), $F_{dB\,res}$ it can be assumed that it is equal to $F_{dB\,RM}$ (see text).

III.7.4 Measuring Reciprocal Mixing

The measuring setup for determining the desensitizing effect of reciprocal mixing is shown in Figure III.55. It is common practice to measure the effect of reciprocal mixing separately for class A1A and J3E emissions only, as described below. Furthermore, some manufacturers provide information (especially in regard to other emission classes) only about the SSB noise of the first LO injection oscillator in a certain frequency separation in order to allow the estimation of the receiver characteristics in the presence of very strong interfering signals.

Measuring procedure:

1. Tune the receiver to the frequency range to be tested.
2. Connect an rms voltmeter or an AF level meter to the AF output of the receiver.

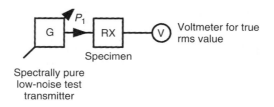

Figure III.55 Measuring arrangement for determining reciprocal mixing.

3. Note the AF output level.
4. Adjust the test transmitter frequency to the frequency separation from the receive frequency at which the reciprocal mixing is to be determined. (A frequency position of 10 kHz or 20 kHz is often used for measurement.)
5. Increase the RF level P_1 of the test transmitter now connected until the entire AF output level rises by 3 dB or the AF output voltage rises to 1.41 times the original value.
6. Note the P_1 level.

The ratio between P_1 and the minimum discernible signal (Section III.4.5) is often called the *blocking dynamic range*. This depends on the receive bandwidth used during measurement to the same degree as the minimum discernible signal. In order to bypass this uncertainty, it is practical to relate the oscillator noise to a fictitious receive bandwidth of 1 Hz. This necessitates a correction of the blocking dynamic range by the equivalent noise bandwidth of the receiving path, expressed in dBHz. This allows the calculation of the SSB noise for the entire receive path according to the relationship

$$L_{RX}(_\Delta f) = P_{MDS}(B_{-6\,dB}) - B_{dB\,N} - P_1(_\Delta f) \qquad \text{(III.26)}$$

where
$$L_{RX}(_\Delta f) = \text{SSB noise ratio of the receiver, depending on the frequency spacing}$$
$$(_\Delta f), \text{ in dBc/Hz}$$
$$P_{MDS}(B_{-6\,dB}) = \text{minimum discernible signal of the receiver with a receive bandwidth}$$
$$(B_{-6\,dB}) \text{ used for measurement, in dBm}$$
$$B_{dB\,N} = \text{equivalent noise bandwidth of the receive bandwidth used, in dBHz}$$
$$P_1(_\Delta f) = \text{signal level of the test transmitter at the receiver input, at which an}$$
$$\text{AF increase of 3 dB is obtained, in dBm}$$

If P_1 is -17 dBm with 50 kHz separation from the receive frequency in a receiver having a minimum discernible signal of -122 dBm at a -6 dB bandwidth of 2.4 kHz, then the oscillator noise for this frequency separation is

$$L_{RX}(_\Delta f = 50 \text{ kHz}) = -122 \text{ dBm} - 33.8 \text{ dBHz} - (-17 \text{ dBm}) = -138.8 \text{ dBc/Hz}$$

This assumes an equivalent noise bandwidth the same as that for the -6 dB receive bandwidth, as is common in most cases for sufficient accuracy. With Equation (III.8), this leads to

$$B_{dB\,N} = 10 \cdot \lg\left(\frac{2{,}400 \text{ Hz}}{1 \text{ Hz}}\right) = 33.8 \text{ dBHz}$$

An analogous procedure is used for determining the spurious component of the LO injection signal. Often, it proves more efficient to continuously tune the receive frequency to values above *and* below the test transmitter frequency (or in the assigned channel pattern) instead of the P_1 frequency. The difference between the levels of P_1 and the minimum discernible signal corresponds to the spurious signal ratio and indicates the range (the dynamic range) in which receiving quality losses are just not yet manifest.

III.7.5   Measuring Problems

With high-quality radio receivers this type of measurement requires a test transmitter with extremely low SSB noise and spurious values. The P_1 signal supplied should be at least 10 dB above the SSB noise of the receiver under test. Otherwise, with a well-designed specimen only the properties of the P_1 signal supplied to the input will be measured! While so far commercially available top-quality test transmitters were sufficient to evaluate reciprocal mixing with adequate accuracy for frequency spacing up to ± 15 kHz, measurements for larger frequency separations can be performed only with low-noise quartz oscillators having a downstream stepping attenuator for level variations. This arrangement is best suited for all measurements of this kind.

With a frequency separation of less than half the first IF bandwidth the effect of blocking (Section III.8) is also present in its true sense. This can be recognized by the fact that P_1 variations produce a decrease in the AF level prior to the 3 dB AF increase. In this situation a meaningful measurement of the reciprocal mixing is no longer possible. This occurs mainly in wideband receivers (so-called scanners) offering the possibility of demodulating broadband-modulated signals with a frequency separation up to 30 kHz and more.

III.8   Blocking

III.8.1   Compression in the RF Frontend or the IF Section

Ideally, in a transmission element the output signal follows the variations of the input voltage. With an amplifier, for example, the output curve will take an identical course with the supplied sinusoidal test tone, but with the amplitude proportionally 'increased' by the gain factor. In reality, signal progressing chains have a restricted control range (dynamic range); with an increasing modulation the signal becomes more distorted until, at the saturation point, it reaches a (sudden) limit. This is the situation of overloading, that is, the gain in the signal path decreases or the attenuation (e.g., of a passive mixer) increases. It is the reason for the curvature in the transfer characteristic (Fig. III.56). The

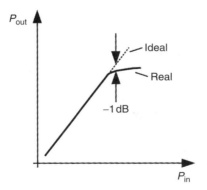

Figure III.56   With too high input signals, the output power no longer rises in proportion to the input power. The transfer element goes into compression.

input level that causes the output signal to fall 1 dB below the output signal expected in case of a straight forward projection, is called the *1 dB compression point*. It is an important parameter for RF assemblies and is used to define the overload point and is often used to specify the maximum dynamic range or linear transmission range.

Care should be taken, if a weak signal is present at the antenna input of the receiver together with a strong signal. Whenever the strong signal (the *blocker*) drives the frontend or the IF stage into saturation, the weak signal, too, is transferred with a respectively reduced gain.

III.8.2   AGC Response to Interfering Signals

The control voltage of the automatic gain control (Section III.14) is derived from the signal mix at the IF output or directly from the audio signal by rectifying and smoothing. It is the mean value of the sum of all signals that have *not* been rejected by the selection. Additive heterodyning of a strong interfering signal with the useful signal in the channel used produces a sum signal. It causes the AGC to reduce the IF amplification and possibly the RF amplification. (The required characteristics of an AGC and its dimensioning are detailed in Section III.14.)

III.8.3   Reduction of Signal-to-Interference Ratio by Blocking

Blocking describes the attenuation of the useful signal by an unmodulated interfering carrier (Fig. III.57). With little frequency separation between the useful signal and the interfering signal the AGC is usually responsible for the blocking effect, while with larger

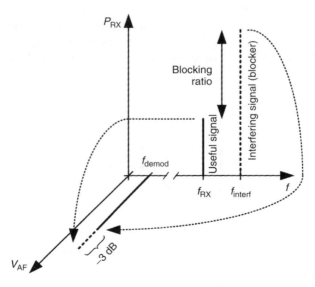

Figure III.57   A (very) strong interfering signal close to the receive frequency 'blocks' the receiver. Due to this blocking effect the volume of the demodulated signal is reduced.

frequency separations it is mostly the result of the compression of the transfer elements caused by the interfering carrier. In reception of class A1A, A3E, and J3E emissions blocking becomes noticeable by a direct volume reduction or by the so-called 'pumping', and with class F3E emission it is the audible SIR decrease, that is, the useful signal sound 'noisier', especially with weaker received signals. When such interferences occur in real receiver operation, then a partial improvement in the reception can be achieved by deactivating the AGC and setting the RF gain and the volume manually (Fig. I.5).

III.8.4 Measuring the Blocking Effect

The measuring setup for determining the volume reduction or the SIR decrease due to blocking is shown in Figure III.58.

Measuring procedure:

1. Tune the receiver to the frequency range to be tested.
2. Tune the test transmitter and with it the receiver to the receive frequency, while observing the attenuation figure of the power combiner. Supply a defined RF level P_{use} and modulate it with nominal modulation (Section III.2). (In amateur radio services often use a P_{use} of -79 dBm [9], while other radio services prefer a P_{use} of -60 dBm or -52 dBm.)
3. Connect an rms voltmeter or AF level meter and SINAD meter to the AF output of the receiver.
4. Increase the RF level P_{interf} of the second test transmitter above the receive frequency until the AF level decreases by 3 dB or the AF output voltage decreases to 1.41 times the initial value, or the SINAD (Section III.4.8) is reduced to 20 dB. (With receive frequency ranges below 30 MHz a frequency spacing of 10 kHz, 20 kHz, 30 kHz or 100 kHz and above it a frequency spacing of 1 MHz is often used for testing.)
5. Note the P_{interf} value.

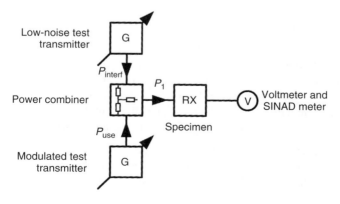

Figure III.58 Measuring arrangement for determining the blocking effect.

The value of P_{interf} corrected by the attenuation figure of the power combiner represents the (absolute) 'blocking value'. The difference between P_{interf} and P_{use} is called the *blocking ratio* (or blocking distance). It varies with the frequency spacing between the useful signal and the interfering signal and with the level of the useful signal.

III.8.5   Measuring Problems

In high-end radio receivers the first indication of blocking is the decrease to 20 dB SINAD caused by reciprocal mixing (Section III.7) in most cases. With F3E the desensitizing effect cannot be determined by observing the specimen. In this case, the test transmitter generating the interfering signal is very important regarding the SSB noise. With a higher frequency separation it is recommendable to use a low-noise Quartz oscillator with a downstream stepping attenuator for level variations. This arrangement is best suited for all measurements of this kind.

When testing receivers for class A3E emissions it can happen that no decrease in volume is noted in the presence of interfering signal levels that are much higher than those causing a reduction to 20 dB SINAD. For the reasons stated above, a (parallel) search for causes of volume reduction *and* SINAD decrease is important.

III.9   Intermodulation

If two or more signals of high levels are present together with the useful signal at the antenna input, their interaction always causes additional synthetic signals that were not present in the signal spectrum before. This is called *intermodulation* (IM). According to the origin, one distinguishes between signals resulting from addition or subtraction of the interfering carrier frequencies and signals grouped immediately above or below the interfering carriers.

All of these signals are called *intermodulation products*. They occur in non-ideal receivers and can have a negative effect on the SIR of signals at the receive frequency, depending on the frequency constellation [26].

III.9.1   Origin of Intermodulation

Limitations or non-linear amplification cause such effects. This is the case if signals present in the existing system exceed the linear range of the transfer characteristic for the system (Fig. III.56). If the gain characteristic of an active component is known, its properties with single-tone and multi-tone control and the resulting distortion products cannot only be measured, but calculated as well. This includes the compression point (Section III.8.1), intermodulation products, and the origin of cross-modulation (Section III.10).

Based on the process of its formation, the relation between these interferences and their interaction can be determined mathematically. This will be rigorously explained with mathematical formulas in the Section V.3, while it is described here verbally for easy comprehension.

III.9.2   Second- and Third-Order Intermodulation

III.9.2.1   IM2

IM2 should be the only intermodulation interference in radio receivers designed in accordance with the latest findings in RF engineering. Unfortunately, this is not the case with the equipment presently used. Particularly at frequencies around 14 MHz and 21 MHz, for example, a large number of IM products can be found. These are *second-order intermodulation products* (IM2). Assuming only two intermodulation interference signals the affected frequencies (Fig. III.59) are

$$f_{IM2} = f_1 + f_2 \qquad\qquad (III.27)$$

$$f_{IM2} = f_2 - f_1 \qquad\qquad (III.28)$$

where

f_{IM2} = frequency of the IM2 components, in Hz
f_1 = frequency of the interfering carrier at the lower frequency, in Hz
f_2 = frequency of the interfering carrier at the higher frequency, in Hz

These result predominantly from strong interfering carriers from SW broadcast regions of lower frequencies (mostly around 7 MHz) reacting with each other. To reject such interferences, receiver manufacturers have divided the entire short wave range into sometimes more than a dozen frequency ranges, each with its own input filters. But particularly the active circuit components used for selecting these bandpasses have been poorly chosen and applied in many cases, so that they produce such interference instead of rejecting it.

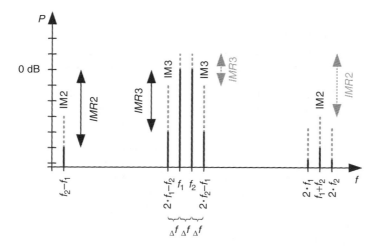

Figure III.59   Frequency position and level response for interference products up to third order. With a continued increase of the excitation signals f_1 and f_2 the intermodulation (distortion) ratio (IMR) decreases rapidly. The level of the second harmonics ($2 \cdot f_1$, $2 \cdot f_2$) always remains 6 dB [21] behind the intermodulation product of second order that evolves in their centre.

From Equations (III.27) and (III.28) it is immediately apparent that these signal products occur at the sum frequency or difference frequency of the two interfering carriers (insofar as we consider the signals at f_1 and f_2 to be interfering signals). Originally, no signals were present at these sum and difference frequencies. They have again been produced by the non-linearity in the transmission characteristic of the receiver. It can also be seen that these interfering products depend on the amplitude of the signal with f_1 and the amplitude of the signal with f_2 in a linear relationship. When increasing one interfering carrier by an amount of x dB, the interfering product increases by the same amount. When increasing both interfering carriers, the interfering product increases by the sum of both in dB! If the two interfering carriers are equal (usually the case in measuring practice), the intermodulation products of the second order respond with a quadratic increase in amplitude or with double the amount in dB (Fig. III.59). This compels receiver developers to implement *input bandpass filters* with *sub-octave bandwidths. Only filters with an upper limiting frequency that is lower than twice the lower limit frequency can prevent the formation of the second harmonic (Section III.5.5) and second-order IM products in the preamplifier and mixer (etc.).*

III.9.2.2   IM3

In modern receivers such as currently marketed, only a few interfering phenomena are dominant. Since new developments for special applications have an RF preselector that narrows the input selection, third-order intermodulation products (IM3) prevail within the receive range. Assuming the existence of only two intermodulation interference signals, the affected frequencies are

$$f_{IM3} = 2 \cdot f_1 - f_2 \qquad (III.29)$$

$$f_{IM3} = 2 \cdot f_2 - f_1 \qquad (III.30)$$

where

f_{IM3} = frequency of the IM3 components, in Hz
f_1 = frequency of the interfering carrier at the lower frequency, in Hz
f_2 = frequency of the interfering carrier at the higher frequency, in Hz

This means that IM3 also occurs with a spacing of $_\Delta f = f_2 - f_1$ below f_1 and above f_2. This is particularly annoying, since it is so close to the critical interfering carriers and can barely be sufficiently suppressed by the usual selection methods (Fig. III.59).

The following model may contribute to a better understanding. A two-tone signal produced by the two interfering carriers corresponds to a beat in the time domain. It can be seen as a double sideband modulated signal (DSB signal) with a suppressed carrier (Fig. III.60). The (suppressed) carrier is modulated with a signal corresponding to half the frequency spacing (f_{mod}) of the two interfering carriers (Fig. III.61). If the signal is cut off when passing the transmission element, as shown in the figures, then odd harmonics of f_{mod} are produced, and with these the IM products in a symmetrical arrangement around the carrier. This limiting effect of the beat comes very close to the real formation process of the *intermodulations of third order*. (The mathematical model, comprising Equations (III.29) and (III.30) with the addition and subtraction of the signals and signal harmonics, is used only for calculating the frequencies at which IM3 products occur.) This suggests:

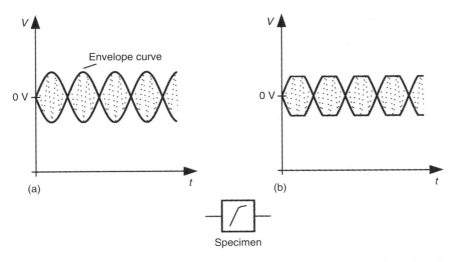

Figure III.60   Illustration of a two-tone beat in the time domain. Graph (a) shows its full amplitude and graph (b) shows the cut off envelope curve after the signal passes a non-linear circuit component.

Short-circuiting the base and emitter of a small-signal amplifier (via a large capacitor for the low differential frequency and decoupled by an RF choke) (Fig. III.62) will improve the intermodulation distortion of the stage by up to 10 dB for a small spacing of the interfering carrier frequencies [26]. In practice, the frequency spacing of the interfering carriers plays a decisive role in the origin of intermodulation! Since usually not even RF

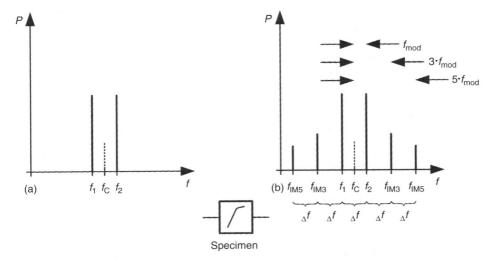

Figure III.61   Illustration of the limiting effect on the formation of intermodulation. The same mechanism is shown as in Figure III.60, but in the time domain. After the limitation the double sideband signal modulated with half the frequency spacing of the interfering carriers cause odd multiples of f_{mod} as shown in graph (b).

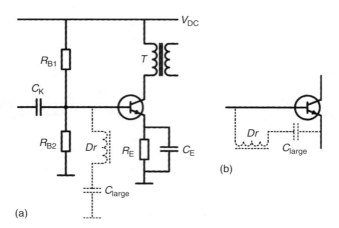

Figure III.62 Method to improve the IM3 properties of amplifier stages for small frequency spacing of the interfering signals, depending on the circuitry used. The collector-base capacitance causes feedback from the output to the base, and the beat produced by the intermodulation effect is short-circuited. When constant interfering carrier spacing can be expected, the resonance of the absorption circuit consisting of Dr and C_{large} is ideally placed at $_\Delta f$.

circuits for very wide frequency bands are capable of processing AF signals below 20 kHz, the IM3 behaviour has a significantly worsening effect if the frequencies of the interfering carriers are very close together. With amplifiers and mixers this can be confirmed easily by metrological methods at any time.

Intermodulation products show no linear increase with an increase in the amplitude of f_1 and f_2, but increase much faster (Fig. III.59). If the two interfering carriers are equal, as is usually the case for practical measurements, the third-order intermodulation products change with the third power of the interfering *voltage*. An increase of the interfering carrier level by 10 dB results in an increase of the IM3 products by 30 dB! And vice versa, a signal decrease by 20 dB due to an attenuator directly before the RF frontend (the typical application of usually switchable attenuators in receivers) causes a decrease of the IM3 products by 60 dB, so that they literally disappear. If, however, the interfering carriers are not equal, such as due to a preselector (Section III.11.1), the amplitude of an IM3 product follows the more distantly separated interfering carrier in a linear mode, but the closer interfering carrier in a quadratic mode. (Figs. III.63, III.64 and III.65). This suggests that preselection is suitable for effectively weakening the IM3 products only if the amplitude of both interfering carriers involved can be reduced. Help is available only from multi-circuit tunable RF preselectors with very narrow bandwidths. However, this means high expenditures in both circuitry and cost.

The third-order intermodulation is often detected in a frequency range below 30 MHz, particularly at around 7 MHz because of the SW broadcast bands, but also at the VHF broadcast band. In many cases, the real intermodulation interferences are generally caused by broadcasting stations, their relay stations, and radar signals with higher frequencies. These simply produce the highest RF signals in the frequency spectrum.

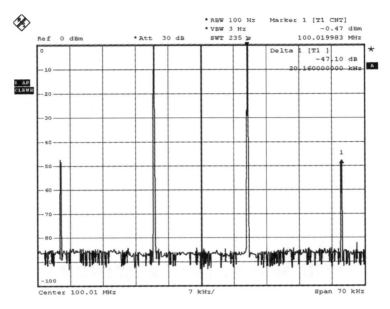

Figure III.63   Output spectrum of an RF amplifier driven by two tones (excitation signals) of the same level.

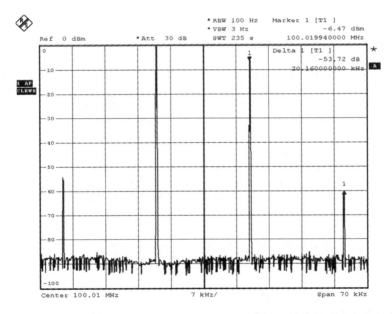

Figure III.64   Compared with Fig. III.63, only the amplitude of the excitation signal of the higher frequency has been reduced by 6 dB. The IM3 product of the lower frequency is clearly stronger than that above the excitation signal.

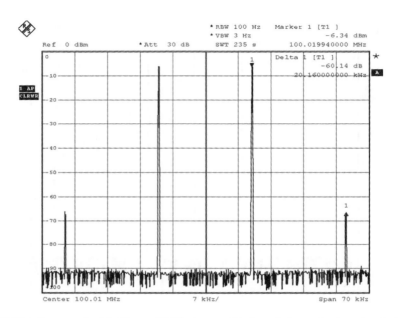

Figure III.65 Compared with Figure III.63, the amplitudes of the excitation signals have been decreased by 6 dB, so that the levels at the specimen are the same again. Here, however, the two IM3 products are reduced significantly and by the same amount (compare with Fig. III.64).

III.9.2.3 Relation between IM2 and IM3

The generation of IM3 in a mixer stage is caused by the speed of polarity reversal in the signal path from the RF port to the IF port due to the oscillator signal. With an increasing slew rate, the IM3 products become smaller. An ideal LO injection signal should therefore be rectangular. The bandwidth compatability of the intermodulation characteristics for a mixer is determined by the choice of ferrite material for broadband transformers and the type of their winding, by the constancy of the LO amplitude, and by the switching characteristics of the diodes (inertia of the charge carriers – switching speed). In practice, very fast Schottky diodes are controlled by a sinusoidal LO injection signal with an amplitude that is controlled and sufficiently high. Even with large frequency variations, the sine wave signal creates a constant relation between the LO frequency and the rate for the diodes and intermodulation properties, which is largely independent of the frequency. Broadband mixers can be dimensioned so that they operate with almost constant data in the frequency range from VLF to the lower UHF band.

In contrast, the formation of IM2 is essentially dependent on the symmetry of the (symmetrical) mixer. This may be independent to a certain degree of the IM3 level. One should therefore view the relation between IM2 properties and IM3 response with caution.

III.9.3 Higher Order Intermodulation

In addition to the frequencies described above and at which intermodulation products occur when the system is controlled by two interfering carriers, there are other frequency combinations that (can) produce intermodulation products.

Although these are of no significance for the interference immunity of the receiver, they may well be the reason for other artefacts. Intermodulation products of higher odd order appear with a very high modulation of a two-port circuit with a frequency spacing of $\Delta f = f_2 - f_1$ above and below the previous IM frequencies of odd order (Fig. III.61). Their influence may cause the sum of the IM3 products to deviate from the cubic level increase (formation law). In other words: Depending on the modulation depth of the tested two-port and its properties (in the case of a diode ring mixer, a MOSFET switching mixer, or a bipolar active mixer) from a certain dynamic range limit on the IM3 products can increase by, for example, 2.5 dB or even 3.5 dB due to a 1 dB rise of the interfering carriers. This deviation can be caused by components of the fifth order. Even the seventh or higher orders can play a theoretical and practical role (Fig. III.66). The following is important for practical work: IM3 products most likely follow the cubic law, provided

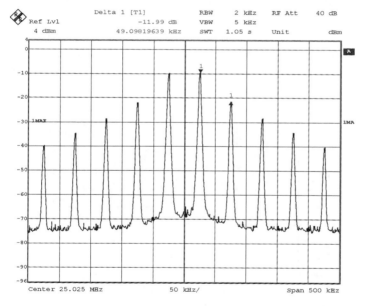

Figure III.66 The graph shows the formation of the maximum achievable IM level for a 7 dBm diode ring mixer. The modulation with two particularly strong excitation signals causes many IM products of odd orders. The excitation signals drive the specimen so far into compression that there is (in this particular case) no further reduction in intermodulation ratio even if the excitation continues to increase. For the IP approximation following the measurement, the specimen should be controlled by levels clearly below the 1 dB compression point.

that the interfering carrier levels are at least 20 dB below the 1 dB compression point (Section III.8.1) or the intermodulation ratio is at least 60 dB.

III.9.4   The Special Case of Electromechanical, Ceramic and Quartz Filters

Just as with active circuit elements, intermodulation can also arise with passive circuit elements. This is called *passive intermodulation*. In this respect, IF selectors play an important role in radio receivers. With both electromechanical filters and ceramic filters, the level increase (in dB) of IM3 products is not always in the proportion of one to three. Rockwell-Collins, the manufacturer of the legendary Collins filters, refers in the current application brochures to this anomaly and states explicitly that 'the slope of the third order products is 2.6 and not 3' (Fig. III.67). In fact, with dual-tone modulation the signal components of IM products of the fifth or seventh order outweigh that of the third order (Section V.3.3). Quartz filters can be even more affected by the 'slow rise' of IM3 products. Sometimes their level increase is only one to two (in dB) in regard to the excitation signals (Fig. III.68). The fact that the increase of IM3 products provoked by the interfering carriers at the input can follow a hysteresis curve and that one cannot always assume a certain intermodulation ratio (IMR) with an alteration of the interfering carrier power is described in [28].

When measuring the intermodulation immunity, it should be noted that with narrow filters the interfering carriers are very close to or already on the filter edge (Fig. III.42). The power combiner in the measuring circuit is then no longer really terminated (by

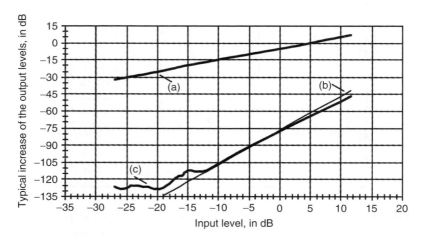

Figure III.67   Typical in-band intermodulation behaviour (Section III.9.12) of electromechanical filters, illustrated for the example of the Collins SSB IF filter like the one shown in Figure III.41. Graph (a) shows the power of one of the supplied excitation signals, and graph (c) shows the real level increase of an IM3 product. With a slope of about 2.6 it does not rise as fast as the ideal line with a slope of 3 shown in graph (b) for comparison. With a weak signal level the response deviates even more.

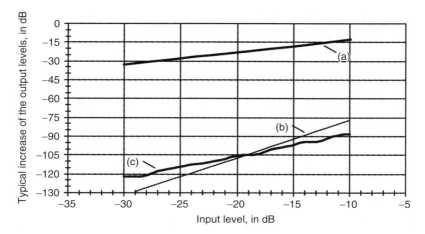

Figure III.68 Typical out-of-band intermodulation behaviour of quartz filters, illustrated for the example of a 45 MHz IF filter according to [27]. The excitation signals are supplied in the cutoff region with 50 kHz spacing. Graph (a) shows the power of one of the supplied excitation signals, and graph (c) shows the real level increase of an IM3 product. With a slope of about 2 it rises clearly slower than the ideal line with a slope of 3 shown in graph (b) for comparison.

a characteristic resistive load) and thus no longer isolates effectively (see Fig. III.80). The same is true for the combination of a mixer and a downstream quartz filter. Even a diplexer is not so narrow that this type of filter provides real termination of the mixer. The only remedy is the insertion of an ohmic attenuator for decoupling and matching by the Pi filter to convert the high-impedance cutoff region of the narrow filters to low-resistive termination impedances for the mixer. Conclusion: Narrow filters in the usual quartz technology do not necessarily have worse intermodulation properties than wider filters, even though this may be suggested repeatedly by incorrect circuit applications or measuring techniques not suited to the situation. Suitable measuring methods are described in [27].

In receiving paths of high large-signal immunity the intermodulation response can be dominated by the IF filter and, with high modulations, show changes according to the generally applied law of cubic increase of IM3 products. Today, it is possible to dimension the individual components and transfer elements of highly linear active RF frontends so that the restricted dynamic capability of available IF filters sometimes defines the limits.

III.9.5 The Special Case of A/D Converted and Digitally Processed Signals

Analog/digital converted signals can also show intermodulation. This depends on the linearity and dynamics of the A/D converter used. Often, these are *not IM interferences in the conventional sense*. The IM products, for example, of the third order can in fact be detected at the frequencies calculated with Equations (III.29) and (III.30), but with an increasing level of the intermodulating excitation signal they may become weaker or even disappear in the noise. The level increase (in dB) of IM3 products in relation to the intermodulation excitation signals simply does not have a slope of one to three (Section V.3.3). This can be explained by the different processing steps of the A/D converter used in accordance with the magnitude of the input signals and the resulting modulation depth. (In other words – the increase or decrease of the IM products is caused by changing the internal serial converter cascades. The switching point depends on the level and follows a hysteresis curve, so that a different response is measured when passing along the curve towards higher or towards lower levels. However, within the dynamic of one of the converter steps the cubic law applies.) The non-linearities are usually spread across the entire range of the A/D converter with a certain periodicity [29]. Theoretically, intermodulation products of the second or higher order can therefore be found across the entire dynamic range.

A remedy is dithering (Fig. I.31), a method for quasi minimizing intermodulation for many input signals to the A/D converter. (In simpler terms, this can be considered to be similar to the effect of the correlative noise rejection method (Section III.2.1), which follows a specific mathematical algorithm.) In an unimportant spectral range a noise signal is added to the useful signal. This significantly reduces by averaging the differential non-linearities (DNL), enabling improvements up to 30 dB [29].

From a technical point of view, quoting intercept points (Section III.9.8) based on approximations as is common for circuits with analog components is not very meaningful (Fig. III.69). More sense makes the intermodulation-limited dynamic range (ILDR)

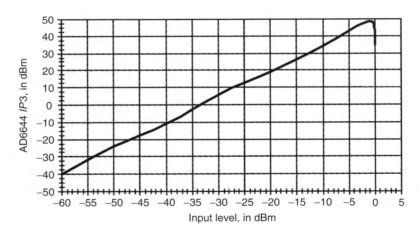

Figure III.69   Variation of $IP3$ shown for the example of the AD6644 A/D converter. It depends on the modulation depth by two intermodulating excitation signals at 7 MHz. While, by definition, the intercept point is independent of the modulation, A/D converters demonstrate an entirely different intermodulation behaviour! (The AD6644 is overloaded with a signal above 0 dB, so that it displays an undefined behaviour. For the values established, the component was operated with a clock frequency of 65.536 MHz.) The specification in the data sheets for A/D converters relates to an $IP3$ value that is usually determined close to the point of overloading (with -7 dBc per excitation signal).

(Section III.9.7), defining the ratio relative to the interfering carrier levels, which remains free of intermodulation interferences. (For the future it is important to find *one* valid measuring procedure. For this, it would be reasonable to define a range up to the point at which a certain IM product level is exceeded for the first time. This would at least include some of the regions in which an initially rising IM product decreases again.) In modern radio receivers (Section I.2.4) that perform A/D conversion and subsequent digital signal processing always closer to or within the RF frontend, the effects explained are often valid for the overall intermodulation behaviour.

III.9.6   Intermodulation Immunity

Intermodulation immunity is the capability of a receiver to prevent the occurrence of interference by intermodulation on the tuned-in receive frequency due to two or more interfering signals at the input having frequencies that interact in some way with the frequency of the useful signal.

III.9.7   Maximum Intermodulation-Limited Dynamic Range

In regard to radio receivers it is interesting to know which level the interfering carriers may have at the antenna input before intermodulation products are induced. This is defined by the maximum *intermodulation-limited dynamic range* (ILDR), also called the intermodulation-free dynamic range. It covers 2/3 of the range between the minimum discernible signal (Section III.4.5) and the relating third-order intercept point. The maximum intermodulation-limited dynamic range (Fig. III.70) depends to the same degree on the measurement of the used receive bandwidth (Section III.6.1) as the minimum discernible signal.

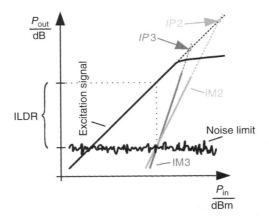

Figure III.70   Extrapolating the ascending curves of the excitation signals of equal power and the intermodulation product in a straight line gives the point of interception at the corresponding IP. Where the IM products remain just below the noise limit, which depends on the receive bandwidth, is the lower edge of ILDR, the maximum intermodulation-limited dynamic range.

III.9.8 Intercept Point

The disproportionate level increase of intermodulation products means that there must be a hypothetical point at which the intermodulation products 'catch up' with the two excitation signals of equal power. The intermodulation products of the respective order would then be of the same power as the provoking interfering carriers. Showing the signal behaviour in a diagram results in the often published presentation of Figure III.70. The point of interception at which the powers have the same value, is called the *intercept point* (IP) of the respective order and is usually expressed in dBm.

In reality, these points of intercept cannot be measured, since the circuitry is driven to its limits by lower input signal. This parameter is, however, still important, because when assuming that the intermodulation products increase as calculated, the IP is independent of the actual modulation. (Comparisons can easily be made even by non-specialists, since this is a single absolute numerical value.) Under these conditions, every measured intermodulation ratio allows the calculation of the corresponding third-order intercept point by the relationship

$$IP3 = \frac{IMR3}{2} + P_{\text{excit}} \tag{III.31}$$

where
$IP3$ = third-order intercept point, in dBm
$IMR3$ = third-order intermodulation ratio, in dB
P_{excit} = level of excitation signals of equal power, in dBm

or the second-order intercept point by

$$IP2 = IMR2 + P_{\text{excit}} \tag{III.32}$$

where
$IP2$ = second-order intercept point, in dBm
$IMR2$ = second-order intermodulation ratio, in dB
P_{excit} = level of excitation signals of equal power, in dBm

With the data deduced from Figure III.63 this leads to the result

$$IP3 = \frac{47.1 \text{ dB}}{2} + (-0.5 \text{ dBm}) = 23.1 \text{ dBm}$$

and with the data deduced from Figure III.65 to

$$IP3 = \frac{60.1 \text{ dB}}{2} + (-6.3 \text{ dBm}) = 23.8 \text{ dBm}$$

Since this is only a different modulation of one and the same transfer component, $IP3$ is hardly influenced. The difference behind the decimal point can be explained by either measuring uncertainties or the reasons stated in Section III.9.3. If the output intercept point determined in this case is to be related to the circuitry input of the RF amplifier under test, the intercept point needs to be corrected by the gain of the component and is then called the input intercept point. In the literature one finds the abbreviations IPIP (*input intercept point*) and OPIP (*output intercept point*) together with its respective order.

Because of the measuring principle with radio receivers, it is always the input intercept point that is determined and stated, so that there is no need for any correction.

III.9.9 Effective Intercept Point (Receiver Factor or ...)

A receiver that is less sensitive, whether owing to the circuitry design or to an attenuation element, usually has a higher intermodulation immunity than a sensitive device – and vice versa. In other words: The *IP3* of a receiver or its noise figure (Section III.4.2) can be increased directly by the attenuation figure of an upstream attenuator, while its sensitivity (Section III.4) is inevitably reduced by the same amount! By inserting an attenuator, the intercept point can be shifted to any desired higher value, which deceives the customer. The maximum intermodulation-limited dynamic range, however, remains the same.

Radio receivers with high intermodulation immunity feed the input signals after preselection (usually with a low-pass filter for image frequency suppression (Section III.5.3), directly to the first mixer without amplification. After the mixer an active or passive intermediate stage follows, to which the first IF filter is coupled. These few elements determine nearly all parameters of the large-signal behaviour (provided that the LO injection signal of the first mixer is of sufficient spectral purity). Such receivers are entirely comparable, since heir noise figure is always within a region of about 12 dB to 16 dB.

In contrast, units for portable use, applications with compromise antennas, and operation with receive frequencies clearly above 30 MHz have a low-noise RF preamplifier directly integrated in the RF frontend. While this guarantees a high sensitivity, it brings a clear degradation for other parameters.

In this respect, receiver systems can vary widely. An objective comparison is hardly possible. The maximum intermodulation-limited dynamic range can be taken as a meaningful criterion for comparison, but there is a problem with differing receive bandwidths (Section III.9.7), rendering a comparison without calculations impracticable. *IP3* should be seen in combination with the noise figure. In [9] and some of the US literature the terms *effective intercept point* or receiver factor and figure of merit (FOM) are also used. $IP3_{eff}$ is determined with

$$IP3_{eff} = IP3 - F_{dB} \qquad\qquad (III.33)$$

where
$IP3_{eff}$ = effective intercept point (third order), in dBm
$IP3$ = third-order intercept point for the radio receiver under test, in dBm
F_{dB} = receiver noise figure, in dB

This shows directly that an upstream attenuator has no influence on the numerical value. The required combination of IM immunity and receiver sensitivity is therefore established without dependency on the bandwidth – *a rather incorruptible criterion for comparison*.

With an RF receiving path having an *IP3* of 17 dBm, determined at 50 kHz spacing from the excitation signal, and a receiver noise figure of 11 dB, the respectable value of

$$IP3_{eff} = 17\ dBm - 11\ dB = 6\ dBm$$

is achieved. The positive value established at the stated frequency separation identifies the device as a good semi-professional unit.

III.9.10   Measuring the Intermodulation Immunity

The measuring setup for determining the intermodulation immunity is shown in Fig. III.71.

Measuring procedure:

1. Tune the receiver to the frequency range to be tested.
2. Connect a SINAD meter to the AF output of the receiver.
3. Tune the test transmitter, and with it the receiver, to the receive frequency, while observing the attenuation figure of the power combiner. Supply the RF level $P_{\text{interf } 2}$, with nominal modulation (Section III.2), so that 12 dB SINAD (Section III.4.8) is obtained at the AF output.
4. Note the value $P_{\text{interf } 2}$, reduced by the attenuation figure of the power combiner.
5. Tune the test transmitter to the interfering tone frequency at the greater separation distance from the receive frequency.
6. Tune the second test transmitter to the interfering tone frequency closer to the receive frequency, which supplies the receiver with an RF level $P_{\text{interf } 1}$ (unmodulated) of equal power.
7. Increase $P_{\text{interf } 1}$ and $P_{\text{interf } 2}$ in parallel by identical RF increments until a SINAD of 12 dB is obtained again.
8. Note the value of $P_{\text{interf } 1}$, reduced by the attenuation figure of the power combiner.

The ratio between $P_{\text{interf } 2}$ in step 4 and $P_{\text{interf } 1}$ in step 8 corresponds to the intermodulation distortion ratio (IMR) achieved with the interfering carrier levels supplied to the antenna socket.

If interfering tone frequencies were selected for determining IM3, *IP3* can be calculated with Equation (III.31). Use the level noted in step 8 for the value of P_{excit}.

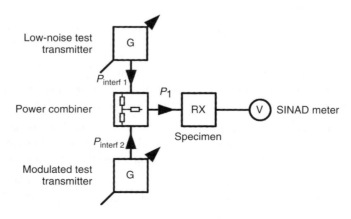

Figure III.71   Basic measuring arrangement for determining intermodulation behaviour.

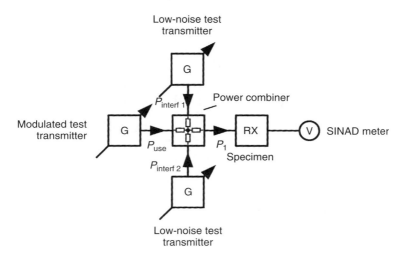

Figure III.72 Measuring arrangement for determining intermodulation behaviour using the so-called three-transmitter method. This procedure is primarily used for testing in compliance with the CEPT ('European Conference of Postal and Telecommunications Administrations') recommendations.

If interfering tone frequencies were selected for determining IM2, *IP2* can be calculated with Equation (III.32). Use the level noted in step 8 for the value of P_{excit}.

Performing measurements in F3E receiving paths with receive frequencies above 30 MHz according to the CEPT test instructions requires the *three-transmitter method* (Fig. III.72). The instructions are followed except for an additional useful signal P_{use} that is permanently modulated with nominal modulation, supplied at the receive frequency. The two interfering signals $P_{\text{interf 1}}$ and $P_{\text{interf 2}}$ are supplied without modulation and increased equally until the demodulated useful AF signal decreases by or below a certain SINAD. If the two interfering signals reach or exceed a predetermined minimum level, the specimen meets the requirements according to the test specifications. The major advantage of the three-transmitter method is that it *also detects* all additional desensitizing effects, like blocking (Section III.8) and reciprocal mixing (Section III.7).

A measuring setup as shown in Figure III.73 has its advantages (see Section III.9.11). When the data determined form the basis of a subsequent *IP3* calculation, the procedure described below is recommended. Despite the initial scepticism regarding possible inter-actions with the AGC (Section III.14), practical measurements have shown that this test is useful for testing short-wave receivers. It has provided useful results in many instances, which has been confirmed by independent sources.

Measuring procedure:

1. Tune the receiver to the frequency range to be tested.
2. Tune the test transmitter, and with it the receiver, to the receive frequency, while observing the attenuation figure of the power combiner. Supply an RF level $P_{\text{interf 2}}$, so that S2 (Section III.18.1) is obtained.

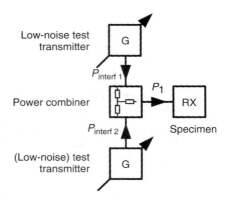

Figure III.73 Measuring arrangement for determining the intermodulation ratio using the S2 method.

3. Note the value $P_{\text{interf }2}$, reduced by the attenuation figure of the power combiner.
4. Tune the test transmitter to the interfering tone frequency at the greater separation distance from the receive frequency.
5. Tune the second test transmitter to the interfering tone frequency closer to the receive frequency, which supplies the receiver with an RF level $P_{\text{interf }1}$ of equal power.
6. Increase $P_{\text{interf }1}$ and $P_{\text{interf }2}$ in parallel by identical RF increments until S2 is obtained again.
7. Note the value of $P_{\text{interf }1}$, reduced by the attenuation figure of the power combiner.

The ratio between $P_{\text{interf }2}$ in step 3 and $P_{\text{interf }1}$ in step 7 corresponds to the intermodulation ratio (IMR) achieved with the interfering carrier levels supplied to the antenna socket.

If interfering tone frequencies were selected for determining IM3, $IP3$ can be calculated with Equation (III.31). Use the level noted in step 7 for the value of P_{excit}.

Warning! Do not change the procedure for S units by using $>$S2, since this would make the interfering signals $P_{\text{interf }1}$ and $P_{\text{interf }2}$ used for the measurement inadvertently too high. This increases the probability of obtaining false results for the reasons discussed in Section III.9.3, since the RF frontend can be driven deep into compression (Section III.8.1). In such a case, the calculated intercept points would be *falsified to the benefit of the specimen*. (It is, therefore, necessary to confirm that the cubic relation between interfering carrier and IM still applies by varying the levels of $P_{\text{interf }1}$ and $P_{\text{interf }2}$ during the measurement.)

III.9.11   Measuring Problems

III.9.11.1   The Problem of SSB Noise

When performing measurements only for gaining information about the intermodulation immunity, one faces the problem (especially when determining IM3) of having to perform tests with interfering signals of frequencies that are close together. This is especially the case with radio receivers of high intermodulation immunity. The noise components due to reciprocal mixing (Section III.7.3) within the receive channel can desensitize the receiving channel so much that smaller intermodulation products are masked. This problem occurs even with radio receivers in which the SSB noise (Section III.7.1) of the LO injection signal in the first mixer stage satisfies the requirements in regard to SSB noise, but the SSB noise of the test transmitters used is too high.

A remedy is the last measuring procedure described in Section III.9.10 (Fig. III.73). While the SSB noise increase is linear in relation to the interfering carrier levels, the IM3 products (in dB) increase three times as fast (in dB). When using the receive signal strength (S2) displayed for determining the strength of the IM product, the IM product clearly exceeds the inherent noise of the radio receiver as it is increased by the SSB noise. It is therefore the strength of the IM product that is determined, and not the SSB noise. Whether the interference effect measured is actually due to intermodulation can be verified by confirming the cubic relation between interference carriers and the IM products.

Ideally, two low-noise oscillators should be used for measuring radio receivers with high intermodulation immunity when interfering tones of small frequency spacing are involved. A downstream stepping attenuator will then vary the signal levels together that have the same power (see Fig. III.78).

III.9.11.2   The Problem of Measuring Frequencies

Testing the IM3 immunity for receive frequencies below 30 MHz is often performed with frequency spacings of 20 kHz, 30 kHz, 50 kHz, 200 kHz, and/or 500 kHz. The smaller the frequency spacing, the stronger are the effects mentioned above. For certain situations, like operating several stations that are geographically close together and on the same frequency band (on ships or during amateur radio contests) examining the IM3 immunity down to frequency spacing as little as 5 kHz can be useful. Such closely adjacent interfering tone frequencies are almost always within the bandwidth of the first IF filter. The IM3 performance of the specimen is then significantly reduced. (In regard to the bandwidth of the first IF stage this would be due to in-band intermodulation.)

For testing the IM3 immunity at receive frequencies above 30 MHz, interfering signals with a frequency spacing of 50 kHz or a separation adapted to the present channel situation of the respecting radio service are frequently used. For receivers used in radio reconnaissance (Section II.4) and ITU monitoring (Section II.4.4), the International Telecommunication Union (ITU) specifies the use of a frequency spacing of 1 MHz [30] in their current recommendations (the ITU Recommendations).

Intermodulation occurs in circuitry chains of radio receivers. These include stages such as the preselector, preamplifier, mixer, mixer termination with filters, and first IF stages. The *intercept point over all stages*, the overall intercept point (Section V.2) of such a chain is influenced by the intercept points of the individual modules and their gain or attenuation. The characteristics of such a chain are never ideal. Each circuit (regardless of how broad its bandwidth is) has limited linearity regarding frequency response, phase response, and interface impedances, even within a limited bandwidth segment. With several stages in series these imperfections inevitably become noticeable. It must be expected that the intermodulation produced is also subject to such influences, independently of the mathematically determined magnitude. This is most obvious from the different levels of IM products on the frequency axis symmetrical to the interfering carriers. Hardly any cascaded system presents a perfect image of symmetric IMs of equal level. Even the simplest system, consisting of a moderate wide-band power combiner and a diode ring mixer, produces asymmetric levels (Fig. III.74). Small changes in the frequency position or frequency spacing of the excitation signals can change the situation, so that an IM product that was previously less pronounced reaches the same or even higher strength than others. The exact cause for such level asymmetries, which are often combined with compensation effects, is very complex. Most recent theoretical findings suggest that they are of thermal origin and may be linked to the 'AM modulation' of the beat. The consequence of this for practical work is that IM products must always be measured on both sides of interference carriers.

For testing IM2 immunity, it is reasonable to use frequency combinations resulting from the position of broadcast bands and their dominating high signal levels. Table III.4 lists examples of frequency combinations below 30 MHz. When choosing test frequencies for IM2 measurements it is important that no harmonic of an excitation signal exists at the

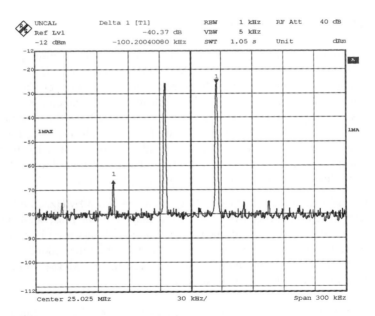

Figure III.74 Different levels of IM3 products due to compression effects in cascaded circuits.

Table III.4 Possible frequency combinations for testing the IM2 immunity of HF receivers

Frequencies of excitation signals	Receive frequency
800 kHz (+) 6.2 MHz	7 MHz
9 MHz (−) 2 MHz	7 MHz
5 MHz (+) 9 MHz	14 MHz
15.2 MHz (−) 1.2 MHz	14 MHz
7.3 MHz (+) 13.7 MHz	21 MHz
9.5 MHz (+) 11.5 MHz	21 MHz

receiving frequency. In such test situations, both the excitation signal frequency and the receive frequency should be slightly shifted to prevent faulty measuring results.

To ensure that the interference is actually an IM product and not a spurious reception (Section III.5) the test transmitters should be turned off alternately.

III.9.11.3　The Problem of Combining Interfering Carriers

Feeding the strong output signal of a test transmitter back to the output can shift the operating point of the output amplifier and produce intermodulation effects. Since very large intermodulation ratios and therefore large differences in signal levels between interfering carriers and the IM product must be measured when testing high-quality radio receivers, the two-tone signal formed by the combination of the two excitation signals must have as little intermodulation as possible.

Consequence: The outputs of the test transmitters must be fully decoupled, if possible. This depends largely on the correct real termination of *all* ports of the power combiner. Especially when measuring receivers with filters at their inputs and therefore having wideband reactive components in their input impedance, decoupling can easily break down (see Figs. III.80 and III.81). An additional ohmig decoupling of 6 dB or more per port by means of a fixed attenuator can prove to be beneficial. In any case, it is essential to deactivate the automatic level control (ALC) of the test transmitters. This is impressively illustrated in Figures III.75, III.76 and III.77.

In receivers with high intermodulation immunity, the accordingly attenuated interfering signals are no longer sufficient for the detection of measurable IM products. The test tones must then be increased by power amplifiers with outputs even more decoupled by attenuators (Fig. III.78). A test measurement with the spectrum analyzer is required to confirm that the power combiner (Figs. III.79, III.80 and III.81) is capable of adding the test transmitter signals with as little intermodulation as possible. This must also be tested in case of insufficient termination or termination with reactive components.

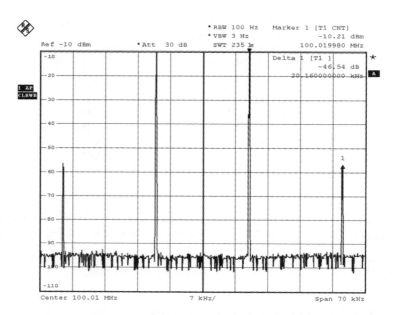

Figure III.75 Owing to interactions between the test transmitters, strong IM3 products become visible after combining the two signals with a power combiner. The spectrum shown was measured directly at the common output of the power combiner on Figure III.71 (resistors in star-connection).

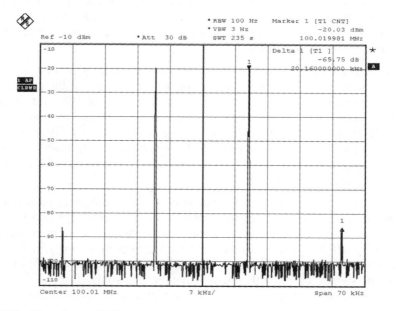

Figure III.76 The often published decoupling of test transmitters by 10 dB attenuators immediately behind the test transmitters leads to an insufficient improvement (compare with Fig. III.75). In addition, valuable test transmitter power is wasted.

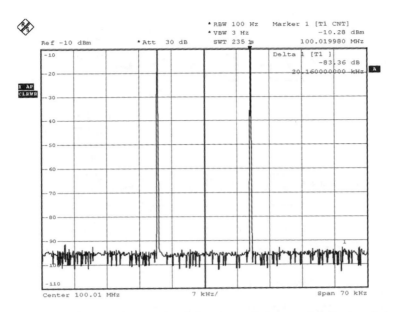

Figure III.77   Only deactivating the ALCs of the test transmitters makes the interfering products disappear completely from the analyzer noise (compare with Fig. III.75). Only this provides a basis for correct measurements.

III.9.12   In-band Intermodulation and Non-Linear Crosstalk

The term in-band intermodulation is generally used for selective measuring objects (their transmission behaviour has bandpass characteristics). For this case, both the excitation signals and the intermodulation products of third order are within the passband region. The measurement of IM3 behaviour in radio receivers often determines the in-band inter-modulation if, in relation to the first IF filter, the interfering carriers are closely adjacent (with spacing of, for example, only 5 kHz).

In J3E receiving paths this is understood to be two interfering carriers within the selected receive bandwidth, which are demodulated together with the resulting IM products to form audible tones at the AF output. (In [28] this is described as an anharmonic interfer-ing component of the output signal caused by the non-linear distortion in the receiver when receiving an input signal of single sideband modulation with a dual-tone sig-nal.) Usually, only the ratio in dB between the demodulated interfering tones and the IM3 products that occur at the reference output power (Fig. III.82) is given. The power of the supplied interfering carrier level is fixed. Besides the term *in-band intermodulation* the expressions inner-band intermodulation, inter-channel intermodulation, and, in legal telecommunication documents, *non-linear crosstalk* are also used.

The correct specification in data sheets is for example: 'Two in-band signals of −23 dBm and 600 Hz spacing produce third-order intermodulation products with a ratio of at least 40 dB.'

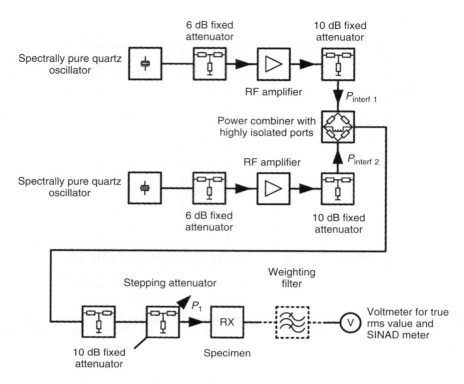

Figure III.78 Required test arrangement for obtaining nearly error-free measuring results for the case of radio receivers with particularly low intermodulation [31]. Owing to the use of quartz oscillators instead of test transmitters, the desensitization of the specimen by the SSB noise (from measuring signals) is minimized.

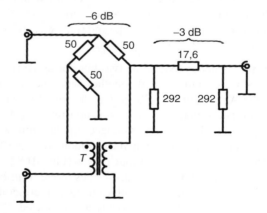

Figure III.79 Principle of a power combiner in bridge-connection [32]. *T* acts as a symmetric transformer. The advantage over a simple ohmic power combiner (resistors in star connection) is the high degree of isolation of the individual ports. To achieve optimum decoupling, another attenuator of >10 dB attenuation is required between the bridge and the measuring object. The symmetry of the bridge remains intact (Fig. III.81) even with a poorly matched receiver input (Section III.3).

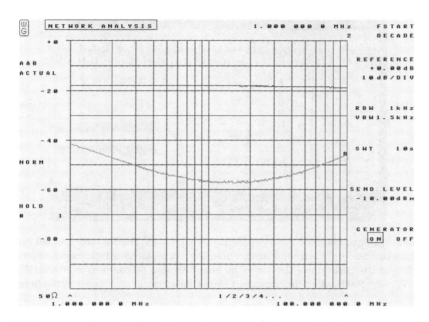

Figure III.80 Decoupling of the two input ports in a sophisticated power combiner in bridge-connection according to Figure III.79. The lower curve results from the real termination of the output port with 50 Ω. If the output is left open (total reflection, refer to Fig. III.81) the decoupling collapses to a level of 18 dB [31].

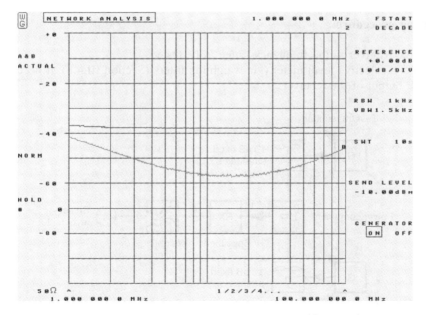

Figure III.81 Decoupling of the two input ports in a sophisticated power combiner in bridge-connection according to Figure III.79. The lower curve results from the real termination of the output port with 50 Ω. However, with an SWR of 1.22 at the termination of the output port (which corresponds to that of a 10 dB attenuator in an open circuit, refer to Fig. III.80) the decoupling is still about 38 dB [31].

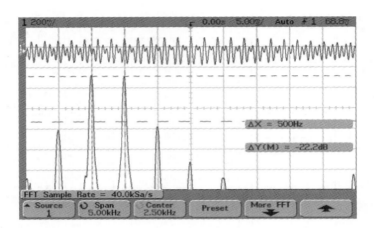

Figure III.82 In-band intermodulation of a J3E receiving path. The AF output signal is displayed. In this particular case, the horizontal marker line has been placed a little higher than the peaks of the IM3 products, since the IMR varies by some dB due to pumping of the AGC. (Y axis: 10 dB/div., X axis: 500 Hz/div.; the upper part shows the demodulated signal in the time domain).

III.9.13 Measurement of the In-band Intermodulation

The measuring setup for determining the non-linear crosstalk is shown in Figure III.83. It is common practice to perform these tests separately only for the reception of class J3E emissions, as described below.

Measuring procedure:

1. Tune the receiver to the frequency range to be tested.
2. Connect a selective level meter (via weighting filters (Section III.4.8), if necessary) to the AF output of the receiver.

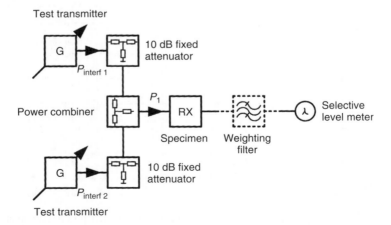

Figure III.83 Measuring arrangement for determining the in-band intermodulation in J3E receiving paths.

3. Tune the test transmitter sufficiently close to the receive frequency so that a 1 kHz tone is generated. This supplies a defined RF level $P_{interf\ 1}$ to the receiver, while taking account of the 10 dB fixed attenuation and the attenuation figure of the power combiner. (At the antenna input socket of the specimen signal levels for $P_{interf\ 1}$ and $P_{interf\ 2}$ of −23 dBm, −13 dBm, or 60 dBµV are often used.)
4. Tune the frequency of the second test transmitter to the frequency spacing required (take account of the lower and upper sidebands). Feed the same defined RF level $P_{interf\ 2}$ to the receiver, while taking account of the 10 dB fixed attenuator and the attenuation figure of the power combiner. (An interfering carrier separation of 600 Hz or 700 Hz is often used for testing.)

The ratio of the separately measured levels of a demodulated interfering carrier from the most powerful IM product corresponds to the suppression of the non-linear crosstalk. Instead of using the selective level meter or voltmeter for the true rms values with an upstream bandpass to filter out the frequency desired, the evaluation can also be performed by FFT analysis. The same result can also be obtained by means of a digital storage oscillograph (Fig. III.82) or dedicated software together with a PC sound card.

When these possibilities are not available, an alternative is to increase $P_{interf\ 1}$ and $P_{interf\ 2}$ until the SINAD for a 1 kHz tone at the AF output is reduced to 20 dB. The second interfering tone originating from $P_{interf\ 2}$ must be suppressed selectively by a high-performance notch filter. This provides another parameter for the comparison if test series are performed with several units.

III.10   Cross-Modulation

With radio receivers of simple design it can happen, especially between the SW broadcast bands, that a very strong AM-modulated emission with a frequency close to the tuned-in receive frequency can superimpose its modulation on the useful signal (another AM emission). The loudspeaker then produces the sound of the selected program and, in addition, the interfering program more quietly in the background. It may also be heard in the normally silent break between words. (This requires that the carrier of the useful signal is present at the receiver, although possibly unmodulated.) This effect is called *cross-modulation*. Since it typically appears with AM signals, its practical importance is limited to certain situations only.

Intermodulation phenomena are sometimes wrongly referred to as cross-modulation.

III.10.1   Generation

With receivers of sufficient intermodulation immunity (Section III.9) usually no cross-modulation effects are detected. Cross-modulation is in fact a specific type of intermodulation of the third order. For this reason, the unwieldy term 'inter-cross-modulation' is also in use. For the generation of cross-modulation only *one* strong modulated signal is needed to affect the reception of the simultaneously existing useful signal (Figs. III.84

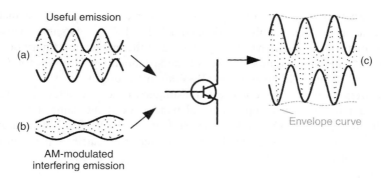

Figure III.84 Cross-modulation exemplified by a transistor stage. Owing to the fact that two signals (a) and (b) are present simultaneously at the base of the transistor the envelope curve carries both the useful signal information and the information of the interfering signal. Due to cross-modulation the envelope of the output signal (c) is also modulated by the interfering signal.

and III.85). If correctly determined *IP3*s are present on a radio receiver the interfering modulation superimposed on the useful signal can be described according to [33] by

$$d \approx \frac{m_{\mathrm{AM}} \cdot 4 \cdot 10^{\frac{P_{\mathrm{interf}}}{10}}}{2 \cdot 10^{\frac{P_{\mathrm{interf}}}{10}} + 10^{\frac{IP3}{10}}} \tag{III.34}$$

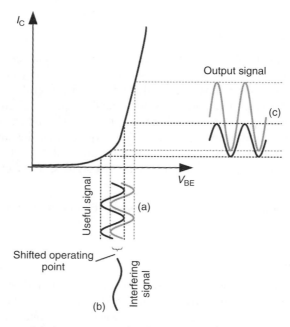

Figure III.85 Cross-modulation process as in Figure III.84, illustrated with the transistor transfer characteristic $I_{\mathrm{C}}(V_{\mathrm{BE}})$. An interfering signal (b) offsets the operating point and causes the gain to vary with time. Even though the useful signal (a) at the base remains the same, its amplitude at the output (c) is influenced by the interfering signal.

where

d = interfering modulation depth superimposed on the useful signal by cross-modulation, in %

m_{AM} = modulation depth of the AM-modulated interfering signal, in %

P_{interf} = level of the AM-modulated interfering signal, in dBm

$IP3$ = intercept point of third order of the receiving path used, in dBm

The parameter determined is actually the so-called interfering modulation depth (or interfering modulation factor). This is the theoretical degree of modulation with which the useful carrier would produce an interfering voltage of the same amplitude as caused by the interfering emission. Other terms, like apparent interfering modulation depth and cross-modulation factor, are also used.

For the following explanation we assume the use of an inexpensively produced radio receiver having an $IP3$ of 5 dBm (determined with the interfering carrier spacing of 50 kHz) and which is operated with a separation of 50 kHz to an MW broadcast emission with an incoming signal of S9 + 60 dB. The modulation depth of AM-modulated broadcast emissions is almost always 30%. According to Table III.6 an S unit of S9 + 60 dB corresponds to an input level of −13 dBm at the antenna socket. Under the given conditions the interfering modulation depth transferred by cross-modulation from the broadcast signal to the useful signal is

$$d \approx \frac{30\% \cdot 4 \cdot 10^{\frac{-13\,\text{dBm}}{10}}}{2 \cdot 10^{\frac{-13\,\text{dBm}}{10}} + 10^{\frac{5\,\text{dBm}}{10}}} \approx 1.8\%$$

This sufficiently low value causes almost no interfering sound from the loudspeaker after demodulation.

An evaluation of the interfering modulation acquired from various strong AM-modulated interfering signals as a function of the intercept point of third order is shown in Figures. III.86 and III.87. It can be seen quite clearly that only very strong interferences affect the quality of reception significantly. If the $IP3$s of a receiving path are above 5 dBm (even with little measuring frequency spacing) there are virtually no interferences due to cross-modulation. Conclusion: With interfering signals of such strength, the adverse effect on the practical receiving situation is more likely caused by blocking (Section III.8) or reciprocal mixing (Section III.7) than by cross-modulation.

III.10.2   Ionospheric Cross-Modulation

The term '*Luxemburg effect*' describes a kind of natural intermodulation. This could also be called *ionospheric cross-modulation*. This phenomenon is not (!) caused by the radio receiver, but by non-linear processes in the ionosphere [34]. It is symptomatic for long-wave and medium-wave ranges. One can imagine that the upward emission of a powerful radio station causes thermal heating of the D layer of the ionosphere at the rhythm of the emitted amplitude modulation.

The name Luxemburg effect dates back to the 1930s. In programs of the Beromuenster radio station operating at a transmit frequency of 556 kHz the program of the Luxemburg

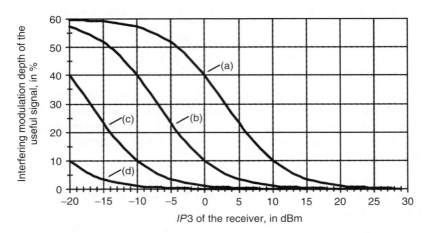

Figure III.86 The interfering modulation depth superimposed on the useful signal is directly related to the intermodulation immunity. The level of the AM-modulated interfering signal is for (a) 0 dBm, (b) −10 dBm, (c) −20 dBm and (d) −30 dBm. All curves are based on a modulation depth of 30%, as is usual in AM broadcasting.

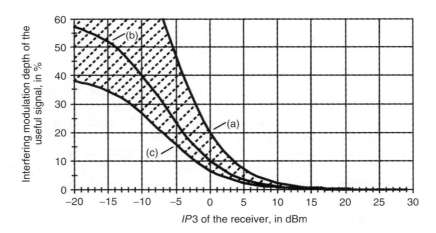

Figure III.87 Influence of different modulation depths of the AM modulated interference signal on the interference modulation depth of the useful signal caused by cross-modulation. Similar to Figure III.86, curve (b) shows an interfering signal of 30% modulation depth. This value has been increased to 60% in (a) and decreased to 20% in (c). The level of the interfering carrier remained constant at −10 dBm.

station could also be heard in receivers tuned to Beromuenster, especially during breaks between words. At the time, the transmit frequency of 230 kHz was assigned to Radio Luxemburg.

Of course, no radio receiver is immune to this kind of ionospheric cross-modulation.

III.10.3 Measuring the Cross-Modulation Immunity

The measuring setup for determining the cross-modulation immunity is shown in Figure III.88. It is common practice to perform such tests separately for class A3E emissions only, as described below. Today, such measurements are often omitted in favour of the more extensive determination of the intermodulation behaviour.

Measuring procedure:

1. Turn the receiver to the frequency range to be tested.
2. Tune the test transmitter to the receive frequency while observing the attenuation figure of the power combiner. Supply a defined RF level P_{use} and modulate it with nominal modulation (Section III.2). (A P_{use} of 60 dBμV or −60 dBm or −52 dBm is often used.)
3. Connect a harmonic distortion meter (via a weighting filter (Section III.4.8) if necessary) to the AF output of the receiver.
4. Subject the RF level P_{interf} of a second test transmitter above the receive frequency to AM-modulation with 400 Hz modulation frequency and a modulation depth of 30%. (A frequency position of 20 kHz or of one or two channel spacings is often used for testing.)
5. Increase P_{interf} until the interfering power due to cross-modulation causes the harmonic distortion at the AF output of the receiver to increase to 3.2%.
6. Note the P_{interf} level.

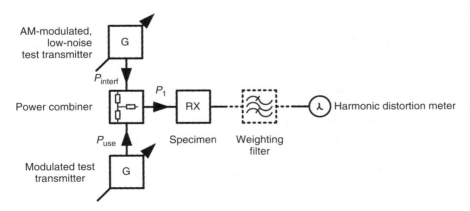

Figure III.88 Measuring arrangement for determining the cross-modulation immunity of receiving paths for class A3E and J3E emissions.

P_{interf}, corrected by the attenuation figure of the power combiner, represents the (absolute) level of an AM-modulated interfering signal that produces an interference ratio of 30 dB below the useful signal at the AF output of the receiver. Under these conditions there is hardly any significant degradation of the reproduction quality (Table III.3). This value varies with the frequency spacing between the useful signal and the interfering signal.

III.10.4 Measuring Problems

Some radio receivers of imperfect circuit design have such a high initial demodulation harmonic distortion (Section III.13.3) and, associated with this, such a low maximum achievable SIR (Section III.4.11) that a worsening of the harmonic distortion by 3.2%, as attempted in the measuring procedure, can hardly be confirmed.

When subjecting a radio receiver of high intermodulation immunity to such a test using an adequate frequency separation between the useful and interfering signals, an increase in the harmonic distortion can also be the result of reciprocal mixing (Section III.7). It is not possible in this case to determine the effect of cross-modulation separately. It is very important that the test transmitter generating the interfering signal has a low inherent SSB noise level. Regarding the combination of useful signals and interfering signals, the considerations of Section III.9.11 should be taken into account accordingly.

III.11 Quality Factor of Selective RF Preselectors under Operating Conditions

Some radio receivers are equipped with *narrow-band preselectors* (also called RF input circuits) arranged directly at the input of the RF frontend behind the antenna input. The preselector serves to relieve the downstream components of the RF frontend from the entire signal scenario provided by the antenna. The signal spectrum around the receive frequency is predominantly passed on. This leads to improved large-signal behaviour (Section III.12), particularly with respect to the strong interfering signals that are not positioned too close to the receive frequency and to the subsequent sum signals.

A number of units of this type operate below 30 MHz. Examples include the Icom IC-7800, Yaesu FT-1000 Mark V, the earlier FT-902 of the same manufacturer and several do-it-yourself concepts of radio amateurs, as well as units with a low first IF to provide relief for the follow-on stages. For the operation of a number of HF transmitters and HF receivers in close geographic proximity (Fig. IV.1) Rohde&Schwarz developed a selector unit as an add-on for the M3SR 4100 equipment family (Fig. III.89), which includes the EK4100 receiver. The design also provides protection against EMF up to 200 V at the unit's antenna socket.

The narrower the filter, the better is the actual receiving performance. The quality of the preselector under operating conditions, that is, the ratio of the receive frequency to its −3 dB bandwidth, is a useful quality parameter. This can change when tuning the centre frequency across the receive frequency range, since sophisticated circuits are required to maintain constant operating quality over the entire frequency range.

Figure III.89   Automatically tracking HF preselector optionally available for upgrading the M3SR 4100 equipment family (Fig. II.10) from Rohde&Schwarz. The unit is available in models with 20 dB or 40 dB stop-band attenuation figure for signals of at least 10% separation relative to the receive frequency. (Company photograph of Rohde&Schwarz.)

III.11.1   Increasing the Dynamic Range by High-Quality Preselection

Evaluating the attenuation of interfering signals separated from the receive frequency can be performed according to [36] by means of the selectivity of a narrow-band RF input circuit tuned to the receive frequency:

$$a_{\text{sel}}(_\Delta f_{\text{interf}}) \approx 20 \cdot \lg \sqrt{1 + Q^2 \cdot \left(\frac{f_{\text{interf}}}{f_0} - \frac{f_0}{f_{\text{interf}}} \right)^2} \qquad \text{(III.35)}$$

where

$a_{\text{sel}}(_\Delta f_{\text{interf}})$ = attenuation figure for interfering signals of the RF preselector circuit, depending on the interfering signal separation $(_\Delta f_{\text{interf}})$, in dB

Q = operating quality of the RF preselector circuit tuned to the receive frequency, dimensionless

f_{interf} = frequency of the interfering signal, in Hz

f_0 = centre frequency of the tuned RF preselector circuit, in Hz

This applies for receiving situations with only one discrete interfering signal affecting the receiving quality (reciprocal mixing (Section III.7), blocking (Section III.8), cross-modulation (Section III.10)) and is similar to the dynamic range increase by selection.

III.11.1.1   IM2

With two intermodulating interfering signals having typical IM2 properties related to the receive frequency according to Equations (III.27) and (III.28) the dynamic range increase by selection is

$$_\triangle IMR2 = a_{\text{sel f}_1} + a_{\text{sel f}_2} \qquad \text{(III.36)}$$

where
$_\triangle IMR2$ = Increase of intermodulation ratio of second order by preselection, in dB
$a_{\text{sel f}_1}$ = attenuation figure of the RF preselector for the interfering carrier of lower frequency, in dB
$a_{\text{sel f}_2}$ = attenuation figure of the RF preselector for the interfering carrier of higher frequency, in dB

of both interfering signals at the input have the same level. (Note: Correct dimensioning of the input band filter with a suboctave bandwidth can ensure sufficient IM2 immunity; see Sections III.9.2 and V.3.1.)

III.11.1.2   IM3

With two intermodulating interfering signals having typical IM3 properties related to the receive frequency according to Equations (III.29) and (III.30) the dynamic range increase by selection is

$$_\triangle IMR3 = 2 \cdot a_{\text{sel f}_n} + a_{\text{sel f}_{fa}} \qquad \text{(III.37)}$$

where
$_\triangle IMR3$ = Increase of intermodulation ratio of third order by preselection, in dB
$a_{\text{sel f}_n}$ = attenuation figure of the RF preselector for an interfering carrier nearer to the receive frequency, in dB
$a_{\text{sel f}_{fa}}$ = attenuation figure of the RF preselector for an interfering carrier further away from the receive frequency, in dB

if both interfering signals at the input have the same level. Unlike with Equations (III.29) and (III.30), Equation (III.37) uses the frequencies f_n (near) and f_{fa} (further away) instead of f_1 and f_2 to meet the conditions for IM3 regarding the receive frequency and the different evaluation of interfering signal combinations by selection.

(For these calculations it is assumed that passive intermodulation of the entire preselection circuit components is negligible.)

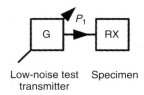

Low-noise test   Specimen
transmitter

Figure III.90   Measuring arrangement for determining the frequency response of RF preselectors.

III.11.2   Measuring the Frequency Response

The measuring setup for determining the frequency response of narrow-band RF prese-lectors is shown in Figure III.90. Measuring this parameter is reasonable only in cases in which the RF preselector can be tuned manually and allows deactivating an auto tracking function if this exists.

Measuring procedure:

1. Tune the receiver to the centre of the frequency band (corresponding to f_0).
2. Tune the test transmitter to the frequency in the centre of the IF passband range.
3. Use the S meter (Section III.18) as an indicator and adjust the RF preselector for the maximum forward signal.
4. Set the RF level P_1 so that S5 (Section III.18.1) is obtained.
5. Increase the output level of the test transmitter by 3 dB.
6. Reduce ($f_{\text{low 3 dB}}$) and increase ($f_{\text{up 3 dB}}$) the receive frequency and test transmitter fre-quency until S5 is reached again. Note the respective frequencies.

For radio receivers without an indication of the receive signal strength, the AF output level of class A1A and J3E emissions can be used for evaluation. In this case it is essential to deactivate the AGC. The quality under operating conditions is then

$$Q = \frac{f_0}{f_{\text{up 3 dB}} - f_{\text{low 3 dB}}} \tag{III.38}$$

where
$\quad Q$ = operating quality of the RF preselector tuned to the input frequency,
$\qquad$ dimensionless
$\quad f_0$ = centre frequency of the tuned RF preselector, in Hz
$f_{\text{up 3 dB}}$ = upper limit frequency of the fixedly tuned RF preselector, in Hz
$f_{\text{low 3 dB}}$ = lower limit frequency of the fixedly tuned RF preselector, in Hz

In order to determine the properties of a preselector for a receiving path (see curve a) in Figure III.91) the preselector has been tuned to 7.050 MHz. The -3 dB frequency values are measured at 6.976 MHz and 7.140 MHz, corresponding to an operating quality of

$$Q = \frac{7{,}050 \text{ kHz}}{7{,}140 \text{ kHz} - 6{,}976 \text{ kHz}} = 43$$

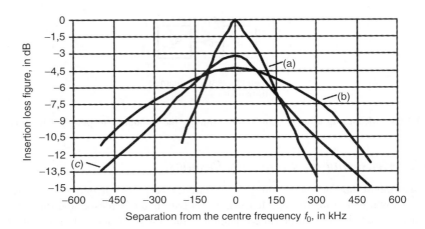

Figure III.91 Estimated attenuation characteristics of different preselectors. (a) shows the curve for a development prototype, and (b) and (c) show the curves for the Yaesu FT-1000 Mark-V Field used in amateur radio service on the 40 m band and the 80 m band, respectively. In (b) and (c) the HF preselector can be deactivated in order to increase the sensitivity. This is not possible with the unit of (a). This is the reason why in (a) the insertion loss figure in the passband has no effect (0 dB).

For interfering signals of 7.2 MHz and 7.35 MHz that lead to the expected IM3 at the receive frequency according to Equation (III.29) with

$$f_{IM3} = 2 \cdot 7{,}200 \text{ kHz} - 7{,}350 \text{ kHz} = 7{,}050 \text{ kHz}$$

according to Equation (III.35) this would lead to an attenuation figure of

$$a_{sel}(\Delta f_{interf} = 150 \text{ kHz}) \approx 20 \cdot \lg \sqrt{1 + 43^2 \cdot \left(\frac{7{,}200 \text{ kHz}}{7{,}050 \text{ kHz}} - \frac{7{,}050 \text{ kHz}}{7{,}200 \text{ kHz}}\right)^2} \approx 6.3 \text{ dB}$$

or

$$a_{sel}(\Delta f_{interf} = 300 \text{ kHz}) \approx 20 \cdot \lg \sqrt{1 + 43^2 \cdot \left(\frac{7{,}350 \text{ kHz}}{7{,}050 \text{ kHz}} - \frac{7{,}050 \text{ kHz}}{7{,}350 \text{ kHz}}\right)^2} \approx 11.4 \text{ dB}$$

in an HF preselector circuit. This compares with the values measured in the dual-circuit HF preselector of Figure III.91(a) of 6.4 dB and 13.9 dB. Using these in Equation (III.37) leads to an improvement of the intermodulation ratio of third order at the receive frequency of

$$\Delta IMR3 = 2 \cdot 6.4 \text{ dB} + 13.9 \text{ dB} = 26.7 \text{ dB}$$

relative to the above interfering signal frequencies and interfering carriers of the same level.

III.12 Large-Signal Behaviour in General

The *real combined action* of all parameters described in Sections III.4 to III.11 can be summarized under one term: *large-signal behaviour. In the professional terminology this defines the behaviour of a radio receiver in a very demanding receiving situation or how it copes with several weak and strong signals of different frequencies present at the antenna input simultaneously.* It is not very meaningful to say that a unit impresses with only one excellent parameter like an LO injection signal with a particularly low SSB noise. Other influencing factors must be equally good and well matched. In a well-designed radio receiver the decisive parameters complement each other and are compatible with each other. By varying the control settings (Table IV.2) the receiving path can often be adjusted to meet specific requirements, like 'particularly high sensitivity' or 'optimum signal-to-interference ratios with strong interfering signals'.

One very demanding situation is that in which several transmitters and receivers are in close geographic proximity, such as aboard ships (Fig. IV.1) or during amateur radio contests. Such operating situations are called collocation. In such situations, reception results of acceptable quality is often possible only with high-quality preselectors of elaborate design (Section III.11) installed directly at the receiver input. Additionally, it may be necessary to use an attenuator (or the input attenuator) to suppress incoming levels in order to satisfy the linearity of the RF frontend within the dynamic range. However, this necessarily causes a decrease in sensitivity (Section III.4).

Of special interest regarding their large-signal immunity are receivers of radio relay stations, since the transmitter and receiver are operating simultaneously in close proximity due to their cross-band duplex operation or duplex mode. This case is characterized by the fact that a single signal of unusual high strength is responsible for the interference, so that intermodulation is not a major issue compared to other parameters. To avoid driving the RF input section into compression (Section III.8.1) owing to the single 'interference' the receiver input is often provided with a high-quality filter at the frequency of the relay station transmitter to selectively attenuate its receive field strength. The considerations outlined above apply accordingly for the other signals at the receiver input not originating in the near vicinity.

III.12.1 Concrete Example

The interaction of the individual parameters will be illustrated for the example of a HF receiving path designed for short-wave reception. A very generously dimensioned wideband antenna system supplying signals with discrete levels of $-30\,$dBm maximum to the input (see also Section III.18.1) is assumed. Furthermore, strong signals as close to the useful channel as $30\,$kHz without having an appreciable effect are assumed. The demodulation is a single sideband demodulation with the rejected carrier, that is, class J3E emission. The receive bandwidth (Section III.6.1) is assumed to be $2.7\,$kHz and, in order to keep other signals well separated from the receive channel, the shape factor (Section III.6.1) obtained with the IF selectors used may not be above 1.8. To enable the radio receiver to pick up weak signals equally well, (besides its large-signal characteristics) a state-of-the-art receiver noise figure (Section III.4.2) of $14\,$dB is assumed. From these key parameters – they could be used as the basis for specifications for equipment development – several operating parameters can be derived. The minimum discernible signal at 2.7 kHz receive bandwidth with an equivalent noise bandwidth according to Equation (III.8) of

$$B_{\mathrm{dB\,N}} = 10 \cdot \lg \left(\frac{2.7 \ \mathrm{kHz}}{1 \ \mathrm{Hz}} \right) = 34.3 \ \mathrm{dBHz}$$

can be calculated with Equation (III.10)

$$P_{\mathrm{MDS}}(B_{-6\,\mathrm{dB}} = 2.7 \ \mathrm{kHz}) = -174 \ \mathrm{dBm/Hz} + 14 \ \mathrm{dB} + 34.3 \ \mathrm{dBHz} = -125.7 \ \mathrm{dBm}$$

In order to prevent the receive channel from being desensitized by more than 3 dB due to the other incoming signals from the air interface, the SSB noise of the LO injection signal caused by reciprocal mixing (Section III.7.3) must be of the same power at most. Therefore, the SSB noise in the receive range of the LO injection signal feeding the (first) mixer stage must be below the level of the $-30 \ \mathrm{dBm}$ strong interfering signal with 30 kHz frequency spacing:

$$L_{\mathrm{B}\,-6\,\mathrm{dB}}(_\Delta f = 30 \ \mathrm{kHz}) = -125.7 \ \mathrm{dBm} - (-30 \ \mathrm{dBm}) = -95.7 \ \mathrm{dBc}$$

Since the calculated value of $-95.7 \ \mathrm{dBc}$ represents the SSB noise ratio in the receive bandwidth, it must be corrected by the already calculated noise bandwidth. This leads to an SSB noise ratio of the LO injection signal with 30 kHz frequency spacing of

$$L_{\mathrm{LO}}(_\Delta f = 30 \ \mathrm{kHz}) = -95.7 \ \mathrm{dBc} - 34.3 \ \mathrm{dBHz} = -130 \ \mathrm{dBc/Hz}$$

To obtain in one step the required SSB noise ratio for the known signal scenario this can be expressed as

$$L_{\mathrm{LO}}(_\Delta f_{\mathrm{interf}}) = P_{\mathrm{MDS}}(B_{-6\,\mathrm{dB}}) - P_{\mathrm{interf}} - B_{\mathrm{dB\,N}} \qquad \text{(III.39)}$$

where
$L_{\mathrm{LO}}(_\Delta f_{\mathrm{interf}})$ = minimum required SSB noise ratio for the LO injection signal, depending on the interfering signal spacing ($_\Delta f_{\mathrm{interf}}$), in dBc/Hz
$P_{\mathrm{MDS}}(B_{-6\,\mathrm{dB}})$ = minimum discernible signal of the receiver for the receive bandwidth used ($B_{-6\,\mathrm{dB}}$), in dBm
P_{interf} = maximum permissible interfering signal level, in dBm
$B_{\mathrm{dB\,N}}$ = equivalent noise bandwidth of the receive bandwidth used, in dBHz

Along with this, the blocking ratio (according to Section III.8.4) relative to a useful signal of $-79 \ \mathrm{dBm}$ must be at least

$$BR(P_{\mathrm{use}} = -79 \ \mathrm{dBm}) = -30 \ \mathrm{dBm} - (-79 \ \mathrm{dBm}) = 49 \ \mathrm{dB}$$

Similarly, no intermodulation product caused by other strong signals via the air interface may affect the reception. Thus, the same radio receiver must have an intermodulation distortion ratio (Fig. III.59) of at least

$$IMR = -30 \ \mathrm{dBm} - (-125.7 \ \mathrm{dBm}) = 95.7 \ \mathrm{dB}$$

in order to ensure that the resulting intermodulation products are not stronger than the noise determining the minimum discernible signal. For third-order intermodulation this IMR can be considered equal to the maximum intermodulation-free dynamic range (Section III.9.7). In accordance with Equation (III.31), the required intercept point of third order (Section III.9.8) is then

$$IP3 = \frac{95.7 \text{ dB}}{2} + (-30 \text{ dBm}) = 17.9 \text{ dBm}$$

According to Equation (III.33) this also corresponds to a required effective intercept point (Section III.9.9) of

$$IP3_{\text{eff}} = 17.9 \text{ dBm} - 14 \text{ dB} = 3.9 \text{ dBm}$$

For the second-order intercept point (Section III.9.8), an intermodulation distortion ratio of the same magnitude also applies. (Note: The excitation signals for second-order intermodulation products are sufficiently separated from the useful frequency band that the excitation signals do not reach the receiver input with full strength due to the selectivity of the antenna or possible matching networks. When taking such conditions into account, the intercept point of second order, which is usually lower than determined with this method, does not necessarily have an adverse effect on the reception quality.) With very wideband assumptions at the receiver input, according to Equation (III.32) the necessary intercept point of second order is:

$$IP2 = 95.7 \text{ dB} + (-30 \text{ dBm}) = 65.7 \text{ dBm}$$

The parameters determined are listed in Table III.5. With a radio receiver of these interacting characteristic parameters a reduction of the signal-to-interference ratio of only 3 dB (due to large-signal effects) must be expected with signals as strong as -30 dBm (corresponding to 14 mV EMF) and have a frequency spacing of ≥ 30 kHz relative to the useful signal! (For receiving conditions in which even higher signal levels exist at the receiver input, a wide-band attenuation pad must be activated directly at the antenna socket in order to keep the HF frontend in the linear segment of the dynamic range.

Table III.5 Interacting parameters of a high-quality HF receiver (see Section III.12.1)

Receiver parameter	Characteristic value
Noise figure	$F_{\text{dB}} = 14 \text{ dB}$
Receive bandwidth	$B_{-6 \text{ dB}} = 2.7 \text{ kHz}$
Minimum discernible signal	$P_{\text{MDS}} = -126 \text{ dBm}$
Reciprocal mixing*	$L_{\text{LO}} = -130 \text{ dBc/Hz}$
Blocking ratio, based on a useful signal of -79 dBm*	$BR = 49 \text{ dB}$
Maximum intermodulation-limited dynamic range*	$\text{ILDR} = 96 \text{ dB}$
Intercept point of third order*	$IP3 = 18 \text{ dBm}$
Effective intercept point*	$IP3_{\text{eff}} = 4 \text{ dBm}$
Intercept point of second order	$IP2 = 66 \text{ dBm}$

*With interfering carrier/carriers with a spacing of at least 30 kHz.

However, this reduces the receiver sensitivity in proportion to the attenuating effect of the attenuator inserted, while the dynamics remain unchanged. Dynamics in this context refers to the level difference between the minimum discernible signal and the highest possible interfering signal that does not yet affect the receiver.)

III.12.2 The IP3 Interpretation Fallacy

An unjustifiably high importance is attached to the *IP3* of radio receivers. Even in the professional literature, the 'all decisive *IP3*' is often glorified. Many users are tempted to believe that one can judge the overall performance of a radio receiver operating in a demanding signal situation from this parameter alone. In fact, there are many restrictions that expressly warn against this!

On the one hand, the limitations described in Section III.9.9 can be remedied by $IP3_{\text{eff}}$. On the other hand, it must be strongly emphasized that *despite any definition, often insufficiently substantiated, there is neither one IP3 nor one IP3$_{\text{eff}}$ for a radio receiver, but – under real conditions – always several values for IP3 or IP3$_{\text{eff}}$.* This is evidenced by numerous test series based on [9] and depends to a high degree on the frequency spacing of the interfering tones causing intermodulation (Section III.9.2.2). The values usually improve with an increasing frequency spacing. How much of an 'improvement' can be expected cannot be predicted from the individual parameter values. To further clarify the situation, Equation (III.31) can be extended to

$$IP3(_\Delta f) = \frac{IMR3(_\Delta f)}{2} + P_{\text{excit}} \tag{III.40}$$

where
  $IP3(_\Delta f)$ = intercept point of third order for a frequency spacing $(_\Delta f)$ of the
          excitation signals, in dBm
  $IMR3(_\Delta f)$ = intermodulation ratio of third order for a frequency spacing $(_\Delta f)$ of the
          excitation signals, in dB
    P_{excit} = level of the excitation signals of equal power, in dBm

Of course, the receive frequency range used or examined within the frequency spectrum covered by wide-band receivers (Section II.4.2) also has an impact.

For example, this effect is very pronounced in some receiver designs for (long-distance) radio telecommunication services below 30 MHz in which automated ohmic attenuators are incorporated behind the antenna input at the expense of the sensitivity (Section III.4.8). This is intended to help avoid operating errors by inexperienced users, but is at the cost of limited flexibility.

The specifications of different manufacturers state IP values or dynamic ranges without any comment about the measuring procedure used. The figures alone may not be considered too meaningful, because of the missing information about the measuring conditions.

Determining the intercept points of radio receivers is particularly critical (Section III.9.10). This requires a procedure tailored to suit the purpose. Even then there are many uncertainties when comparing values from different sources. In this context, it is necessary to cite a sentence often found in the literature in this or a similar wording. It helps to close the (vicious) circle and returns to the statements made at the beginning of this section: 'One may claim that the intercept point of third order is the most important single specification for the dynamic properties, since it shows the performance characteristics regarding intermodulation, cross-modulation and desensitization due to congesting effects.' It would be nice if this were the case. Regarding desensitization and digestion, the statement refers to very particular methods used in determining intermodulation and then only to the intermodulation-free dynamic range! In many cases, the *IP*3 calculated on this basis cannot be taken as such. Instead, it often represents a cryptic number based on incorrect assumptions.

III.13   Audio Reproduction Properties

The type and quality of reproduced AF signals demodulated from a useful RF signal with a high signal-to-interference ratio are generally referred to as the *audio reproduction properties* of a radio receiver. The intensity of the volume achievable is expressed indirectly by the parameter of the specified AF power together with the statement of the nominal load (Equation (III.1)).

III.13.1   AF Frequency Response

The sound is strongly influenced by the intensity of the single spectral components of the demodulated AF signal that reaches the loudspeaker or a unit (like a decoder) used for further processing of the AF signal.

Besides cases consciously using preemphasis/de-emphasis (Section III.4.9), the greatest possible linearity of the AF frequency response is desirable in most cases. All relevant spectral components of the demodulated AF signal should be available at the AF output or at the loudspeaker with equally high levels. While the entire audible range of VHF broadcast radio is between 20 Hz and at least 16 kHz, receivers designed for voice radio are dimensioned so that only the dominant voice frequency components between 300 Hz and 3 kHz are reproduced (refer to Section III.6.1). There is a sharp drop towards the lower or higher frequencies, so that other frequencies are cut off. A well-designed radio receiver without de-emphasis of the AF output level should reproduce the signals of different modulation frequencies within a given frequency range with not more than $+1.5\,dB$ to $-3\,dB$ maximum variation compared with the AF output level of signals with nominal modulation (Section III.2).

For the demodulation of class A1A and J3E emissions, the AF frequency response therefore represents the variation of the AF output level as a function of the sideband frequencies for a constant RF input level.

Under certain receiving conditions, intentionally cutting the AF frequency response can be useful in order to minimize noise components and thus increase the SNR (Section III.4.8). For class A1A emissions a so-called CW pitch filter is sometimes used to forward only the demodulated CW tone in a narrow passband. For digital modulation methods, like audio frequency shift keying (AFSK) used in class F2D emission, defined AF frequency response characteristics are common. These only allow the passage of the two expected discrete tones which define mark and space or high and low. Notch filters are used to reject discrete interfering tones by attenuating part of the AF signals within the passband region. With a notch filter of high quality the sound character can be largely maintained, while the SIR is significantly improved in regard to discrete interfering tones included in the demodulated signal. In addition to the quality factor of a notch filter, its depth which specifies the achievable attenuation figure of the interfering tone, is an important parameter for estimating the efficiency.

III.13.2   Measuring the AF Frequency Response

The measuring setup for determining the AF frequency response is shown in Figure III.92.

Measuring procedure:

1. Tune the receiver to the frequency range to be tested.
2. Connect the voltmeter for the true rms value or the AF level meter to the AF output of the receiver.
3. Tune the test transmitter to the receive frequency and in this way feed an RF level P_1 of -73 dBm with nominal modulation for frequencies below 30 MHz or an RF level P_1 of -93 dBm for frequencies above 30 MHz to the receiver. (Professional radio services often use a different RF level P_1 with an intensity of 20 dB above the operational sensitivity (Section III.4.8).)
4. Note the AF output level.
5. Vary the modulation frequency of the test transmitter and note the resulting changes of the AF output level with nominal modulation. For class A1A and J3E emissions vary the receive frequency at the specimen instead of the modulation frequency with a constant RF input level, resulting in a change of the demodulated AF tone.

When applicable, the complete AF frequency response curve can be shown or evaluated in a diagram, using the AF output level as a function of the modulation frequency.

III.13.3   Reproduction Quality and Distortions

In audio technology the harmonic distortion is preferred as an evaluation factor over the maximum signal-to-interference ratio (Section III.4.11). This allows the qualitative

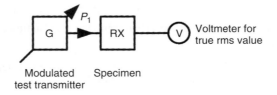

Figure III.92   Measuring arrangement for determining the AF frequency response.

evaluation of a (sinusoidal) signal. This assumes that all harmonics arising in addition to the fundamental wave cause a degradation of the signal quality. According to the definition, the correct mathematical expression is

$$THD_\% = \frac{\sqrt{V_{2.\,HW}^2 + V_{3.\,HW}^2 + V_{4.\,HW}^2 + \cdots + V_{n.\,HW}^2}}{V_{tot}} \cdot 100\% \qquad \text{(III.41)}$$

where
$THD_\%$ = total harmonic distortion, in %
$V_{2.\,HW}$ = effective value of the 2nd harmonic, in V
$V_{3.\,HW}$ = effective value of the 3rd harmonic, in V
$V_{4.\,HW}$ = effective value of the 4th harmonic, in V
$V_{n.\,HW}$ = effective value of the nth harmonic, in V
V_{tot} = effective value of the total signal, in V

The harmonic distortion indicates the ratio of the effective values of the harmonics (without the fundamental wave) to the effective value of the total signal V_{tot}. It is commonly expressed in per cent (Fig. III.93) and is called the total harmonic distortion (THD) [22]. A value of $THD_\% \leq 10\%$ indicates that the sum of the harmonics of the signal is not more than 10% of the total signal. A useful signal with a frequency of 1 kHz is used almost exclusively for measurements in AF technology. This is the modulation frequency of the nominal modulation.

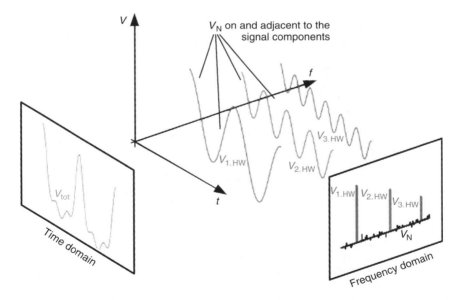

Figure III.93 Typical distortions of a demodulated signal divided into its signal components. The designations of the individual voltage components correspond to those given in Equations (III.12) and (III.41).

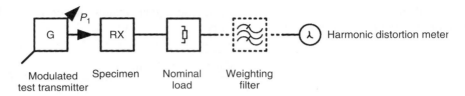

Figure III.94   Measuring arrangement for determining the demodulation harmonic distortion.

An important criterion for the evaluation of radio receivers is the *(total) demodulation harmonic distortion*. This serves as the reference output power for a certain or for several predetermined RF levels supplied to the antenna socket (see Figs. III.95 and II.96). The control range (drive) of the AF output with reference output power as specified in most data sheets of radio receivers is of particular importance, since distortions change with the drive, resulting in variations of the reproduction quality. With sufficiently high input signals a well-designed receiving path can achieve a demodulation total harmonic distortion below 1%. Such low interfering spectral components are hardly audible. The explanations given in Section III.4.11 apply accordingly to the demodulation harmonic distortion.

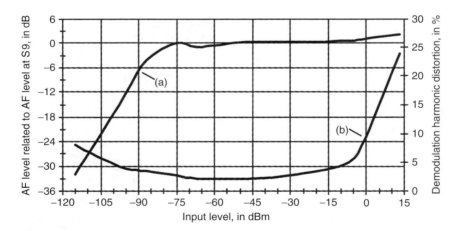

Figure III.95   (a) Shows the AF output signal as a function of the useful signal level supplied to the antenna socket, (b) represents the harmonic distortion of the demodulated signal. With weak input signals the AF signal decreases rapidly. However, from −90 dBm on it remains almost constant at 6 dB. The J3E radio receiver under test does not have a particularly linear demodulation characteristic and is completely overloaded with high input signals. This follows from the steep increase of the demodulation harmonic distortion with RF input levels exceeding −25 dBm.

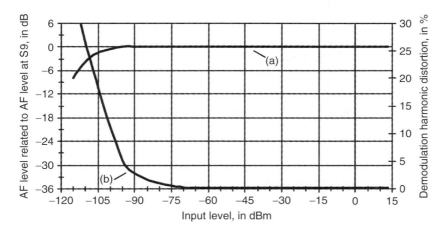

Figure III.96 A well-designed AGC produces a constant output signal as (a) shows. (Compare with Fig. III.95.) Even for very low input signals the volume drops by hardly more than 6 dB. With this J3E radio receiver the demodulation harmonic distortion (b) for sufficiently high RF input signals is below 0.2% and remains at this value even for very strong signals. The fact that in this case the demodulation harmonic distortion with low input signals is not as good as in Figure III.95 is due to the lower receiver sensitivity. However, this has a positive effect with respect to its overload immunity.

III.13.4 Measuring the Demodulation Harmonic Distortion

The measuring setup for determining the demodulation harmonic distortion is shown in Figure III.94.

Measuring procedure:

1. Tune the receiver to the frequency range to be tested.
2. Connect the nominal ohmic load as specified in the specifications of the receiver manufacturer (e.g., 4 Ω, 8 Ω, for headphones 600 Ω) to the AF output of the receiver.
3. Connect a harmonic distortion meter (via a weighting filter (Section III.4.8) if necessary) to the AF output of the receiver.
4. Tune the test transmitter to the receive frequency and in this way feed an RF level P_1 of −73 dBm with nominal modulation for frequencies below 30 MHz or an RF level P_1 of −93 dBm for frequencies above 30 MHz to the receiver. (Professional radio services often use a different RF level P_1 with an intensity of 20 dB above the operational sensitivity (Section III.4.8). Some measuring specifications require determining the demodulation harmonic distortion at several predetermined RF levels P_1 of different intensity.)
5. The display of the harmonic distortion meter provides a direct reading of the demodulation harmonic distortion.

III.13.5 Measuring Problems

In order to obtain reliable measuring results over the AF range of interest the voltmeter or the AF level meter used must have a linear frequency response and a sufficient bandwidth. If the characteristics of a pitch/notch filter are to be determined together with the AF frequency response, the modulation frequency should be adjusted within the passband or cutoff range in very fine increments to accurately determine the maximum/minimum.

When measuring the demodulation harmonic distortion the modulation signals must be spectrally pure, since otherwise the results achieved with radio receivers of good demodulation properties contain an error to the *disadvantage* of the receiver.

III.14 Behaviour of the Automatic Gain Control (AGC)

With emission classes (or modulation types) transmitting the information content through amplitude variations of the RF signal, a control loop regulates the demodulated output signal to a preferably constant AF output signal, independently of the strength of the RF input signal. Without such an automatic gain control the AF signal increases with an increasing RF input signal until the signal processing components are overloaded (Fig. I.5). The *automatic gain control* (AGC) changes the amplification in the various stages in response to the receive signal strength.

For sound reproduction and also for the further processing of the useful signal in downstream units (like decoders) a volume remaining as constant as possible is desirable. As with every control loop certain conditions, like sudden variations of the RF input signal, affect the output signal. One reason is that level changes cannot be counteracted with unlimited speed. Another is that it simply takes some time to 'smooth out' the resulting transient oscillations to the setpoint value (the volume desired). The overall control behaviour of a radio receiver can therefore be divided into static and time dynamic behaviour.

III.14.1 Static Control Behaviour

The *static control behaviour* defines how well the AF output signal can be kept constant with RF inputs ranging from low to relatively high signal intensities. The system is in a steady state. Figures III.95 and III.96 illustrate the typical responses in J3E receiving paths with different types of AGC circuitry.

With very low input signals the overall gain of a receiver is usually not sufficient to bring the AF signal up to the desired preset volume. The difference between RF input levels of 0 dBm and the power level causing the AF signal to decrease by 6 dB (sometimes to 2 dB or 3 dB only) is called the *nominal AGC range*. (This assumes that controlling the gain of signals >0 dB is practically irrelevant.) The point at which the 6 dB AF reduction is reached is called the AGC threshold (the AGC knee or limiting point). The range above is the gain limiting range. With present circuit designs it is possible to cover an AGC control range of more than 110 dB. (As a consequence, the point of 6 dB AF reduction is then below −110 dBm and thus relevant for narrow-band emission classes.)

In fact, the AGC of a modern unit for semi-professional applications cuts in when the signal exceeds S4 or S5 (Section III.18.1) (Fig. III.95). Why? Apart from design deficiencies, the reason is in part a concession to older operators. Some of the older units still

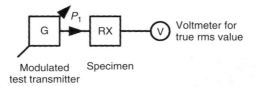

Modulated   Specimen
test transmitter

Figure III.97   Measuring arrangement for determining the static control behaviour of the AGC.

in use have been designed for the audible sound reception of Morse telegraphy or voice radio in class A1A or J3E emissions. (This group of people is used to the peculiarities of units with electron tube, with not more than 60 dB AGC control range. They have used these for several years and always coped with weak input signals by increasing the volume by 20 dB. If this is omitted, the background sounds 'very quiet', since the troublesome background noise cannot be heard.) Radio receivers of this type require manual adjustment of the volume when receiving low input signals.

III.14.2   Measuring the Static Control Behaviour

The measuring setup for determining the static control behaviour of the automatic gain control is shown in Fig. III.97.

Measuring procedure:

1. Tune the receiver to the frequency range to be tested.
2. Tune the test transmitter to the receive frequency and in this way feed an RF level P_1 of 0 dBm with nominal modulation (Section III.2) to the receiver.
3. Connect the voltmeter for the true rms value or the AF level meter to the AF output of the receiver.
4. Reduce the RF level P_1 of the test transceiver until the total AF output level drops by 6 dBm or the AF output voltage decreases to half its value. (For evaluation, test specifications often stipulate the AF drop by 3 dB, that is, 1.41 times the original voltage or 2 dB, equal to 1.26 times the original value instead of the 6 dB decrease.)
5. Note the value of P_1.

The value of RF level P_1 corresponds to the nominal AGC control range of the radio receiver.

III.14.3   Time-Dynamic Control Behaviour

The *time-dynamic control behaviour* is determined by supplying defined input signal level jumps while observing the time the AF output signal takes to approach the setpoint value – the transient settling time – and its oscillation pattern. For the practical operation of a radio receiver there is always the risk that a suddenly incoming RF input signal produces annoying popping noises in the loudspeaker due to strong overshoots (Fig. III.98), such as in two-way radio telephony. This effect is reinforced when using especially narrow IF bandwidths. Another important factor is the time the AGC requires to respond to the sudden stop of an RF input signal. This requires a certain hold time in order to

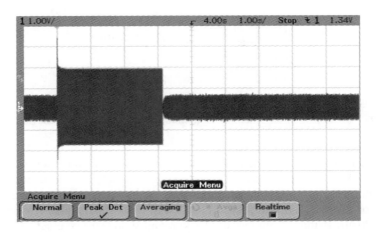

Figure III.98 Time-dynamic response of the AF output signal of a VLF/HF radio receiver set to 'AGC fast' for demodulating class A1A or J3E emissions. The initial marked increase of the AF signal follows a jump in the RF input signal by 30 dB (from −103 dBm to −73 dBm). The decrease in the AF signal following about 3 s later causes an abrupt reduction of the RF input signal by 30 dB (from −73 dBm to −103 dBm). The equipment tested produces an unsettled hearing impression due to the very short time in which the AF signal reaches the final value (that is, the transient oscillation). This setting of the AGC often produces crackling noises with high jumps of the RF input level.

prevent a sudden and irritating surge of the noise for class A1A emissions or in pauses between words of other modulation types without continuous carrier. In sophisticated radio receivers it is usually possible to choose between several AGC time constants in order to achieve an optimum compromise for the emission class used and the personal preferences of the user (Fig. III.99).

The demands that must be met by an automatic gain control for various modulation types are summarized in [6] as follows:

- With class A3E emissions the AGC must react to the sum signal of the carrier and the side bands with some time delay to avoid influences due to variations of the modulation envelope curve. The AGC time constant must therefore be matched to the lowest transmitted or demodulated modulation frequency (e.g., 300 Hz in voice communication or 50 Hz in radio broadcasting). At the same time it must respond fast enough to optimally compensate fading.
- With class A1A and J3E emissions control must be based on the peak value alone and not on a constant level. For this purpose the radio receiver should have an attack time as short as possible after the signal starts. (A general rule is: 1 kHz bandwidth requires about 1 ms delay.) In keying or speech pauses the hold time must prevent the AGC from following the rhythm of the modulation. (In many units the hold time is too short for clean SSB reproduction.) The decay time following the hold time must block the abrupt increase, but follow any fading of the receive signal sufficiently fast (Fig. III.100).

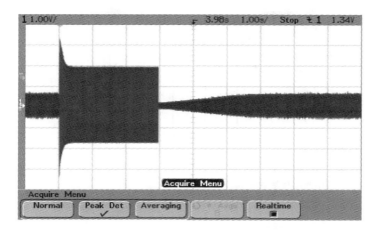

Figure III.99 Time-dynamic response of the AF output signal of a VLF/HF radio receiver set to 'AGC slow' for demodulating class A1A or J3E emissions. This graph was taken under conditions identical to those of Figure III.98, but with a different AGC time. With this AGC setting the equipment tested produces a sound characteristic with a much quieter background.

- With class F3E emission no IF control is required, since the amplitude of the RF input signal carries no information and the IF amplifier limits the amplitude. High-quality F3E receiver paths use such limiting IF amplifiers without any control action exclusively. Older circuit designs and inexpensive equipment for semi-professional application still incorporate RF amplitude control as a protection against strong interfering signals – even though this is in fact harmful. However, this allows the simple indication of the relative signal strength (Section III.18) derived from the control voltage. (An indication obtained in this way shows only limited display dynamics.) The time constants of the control action can be rather short.

The so-called AF AGC is often said to have very good characteristics, since it uses a control voltage according to the overall selection of the receive path. This is offset however by the serious disadvantage of a very slow response, which causes distinctly audible overshoots or limitation effects after data blocks or speech pauses. A truly professional solution is an AGC that responds to the useful signal from the IF and decreases with the AF. This shows very little reaction to interferences (see also Section III.8.2).

III.14.4 Measuring the Time-Dynamic Control Behaviour

The measuring setup for determining the time-dynamic control behaviour of an automatic gain control is shown in Figure III.101.

Measuring procedure:

1. Tune the receiver to the frequency range to be tested.
2. Tune the test transmitter to the receive frequency, modulate with nominal modulation and use a clock generator to trigger the test transmitter by V_{cl} with a frequency of less

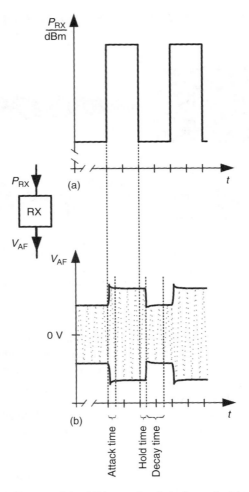

Figure III.100 With level jumps of the RF input signal as shown in (a), the AF output voltage as shown in (b) is not only kept constant over long periods, but changes are intentionally delayed. The times marked in the diagram characterize the time-dynamic control behaviour of a radio receiver and must be adapted to the respective emission class. In practice, an overshooting AF output signal as a consequence of a jump in the RF input signal is normal. With special designs of the AGC one attempts to counteract this behaviour.

than 0.3 Hz, so that the RF level P_1 switches periodically between the lowest and the highest test value. (For the test, adjust RF level jumps of P_1 to values between 20 dB and 80 dB.)

3. Connect a storage oscilloscope to the AF output of the receiver, trigger with the clock generator signal and record the AF output signal with a slow time base.

Evaluate the oscillogram according to Figure III.100. The end of each period determines the point in time when the AF output signal reaches 10% or 90% of the steady state

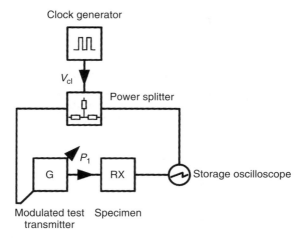

Clock generator

V_{cl}

Power splitter

P_1

G   RX   Storage oscilloscope

Modulated test   Specimen
transmitter

Figure III.101   Measuring arrangement for determining the time-dynamic control behaviour of the automatic gain control.

voltage. Owing to the required slow time base of the storage oscilloscope, the display mode must be set to 'Normal' instead of the automatic preset 'free run' mode.

If a test transmitter with a variable RF output signal and triggered by an external signal is not available, an alternative possibility is to use a relay for short-circuiting the attenuator inserted between the test transmitter output and the receiver input with the clock generator pulse. In this case, precautions must be taken to ensure that the various components or signals are combined with the correct characteristic impedance. Any bounce of the relay contacts must be prevented using appropriate circuit measures.

III.15   Long-Term Frequency Stability

The *long-term stability* (frequency stability) of a receiver is determined by its method of frequency generation. This determines how reliable the adjustment of the receive frequency remains after tuning (or after selecting the frequency channel desired) and how stable it remains over a longer period. The internal oscillators for frequency generation settle on reaching the operating temperature. Only then, after a certain time, is the frequency accuracy within the specified tolerances (Fig. III.102). Nowadays, this warm-up time ranges between a few minutes and half an hour. *Important for this is the so-called mother oscillator* (mostly oscillating with 1 MHz or 10 MHz), from which all other frequencies required internally in a modern receiver concept are derived. Following the warm-up time, frequencies can vary over time due to physically unavoidable aging processes of the elements decisive for the frequency (usually the quartz crystal of the mother oscillator). A typical specification of frequency instabilities due to aging is

$$\pm 5 \cdot 10^{-8}/\text{year},$$

which means that for a frequency adjusted to 1.2 GHz a maximum absolute deviation of 60 Hz is to be expected after one year. The ambient temperature also has an effect on

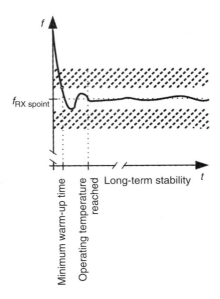

Figure III.102 The specified frequency accuracy is obtained only after the warm-up time, and remains stable when the operating temperature has been reached. Then, the deviation due to drift is within tolerance.

the frequency drift. The deviation in frequency is corrected by the automatic frequency control (AFC) to the set-point frequency with a tolerance depending on the type of auto-tracking or auto-control. A typical specification of the frequency drift due to temperature changes is

$$\pm 2 \cdot 10^{-9}/\mathrm{K},$$

which means that for the assumed frequency of 1.2 GHz a maximum absolute deviation of 2.4 Hz per (degree) Kelvin temperature change. In practice, temperature change and aging occur simultaneously, so that constructive and destructive effects are combined.

The long-term stability also determines how well the set-point receive frequency can be reached and maintained. This is particularly important for search receivers and/or when working with narrow-band emission classes or using narrow receive bandwidths (Section III.6.1), since with excessively large deviations the useful signal desired can, in extreme cases, lie outside the receive bandwidth. With class A1A emissions the annoying frequency drift becomes noticeable in the form of a demodulated tone varying over time. With class J3E/A3E reception the demodulated voice frequency band appears shifted and sounds unnaturally dull or bright. With digital emissions the bit error rate (Section III.4) increases with the deviation from the set receive frequency.

III.15.1 Measuring the Long-Term Frequency Stability

The measuring setup for determining the warm-up behaviour and the long-term frequency stability is shown in Figure III.103.

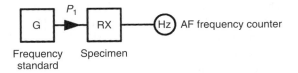

Figure III.103 Measuring arrangement for determining the long-term frequency stability with class A1A and J3E emissions.

Measuring procedure:

1. Tune the receiver to the frequency of the frequency standard so that a tone is generated with the frequency of the BFO offset with class A1A emissions or has a demodulated frequency of 1 kHz with J3E emissions.
2. Connect an AF frequency counter to the AF output of the receiver.
3. Note the frequency indicated on the radio receiver.
4. Note the frequency of the demodulated AF output signal in small time increments over the measuring period.

The difference between the frequency of the frequency standard and the frequency indicated on the radio receiver represents the absolute deviation after its correction for class J3E emissions by the pitch of the demodulated tone. The long-term frequency stability can be determined directly from the frequency variations of the demodulated AF tone. For graphical evaluation it may be helpful to enter the values in a diagram similar to that shown in Figure III.102.

To achieve very high accuracy, it is possible to synchronize the frequency standard, for example, via DCF77 (Section II.7) for continuous signal correction.

III.15.2 Measuring Problems

The frequency counter used for the measurement should have a tolerance exceeding the deviations expected with the specimen by at least one power of ten. Important for the frequency counter is a *sufficiently long warm-up time* before starting the measurement. For the measurements the frequency counter should be placed in a closed room with constant temperature and minimum air convection. Prior to the measurement the radio receiver can be cooled down for a prolonged period in order to obtain the best approximation to the real warm-up behaviour (under extreme conditions).

With class A3E and F3E emissions the measurement of frequency stability with the demodulated RF output signal as described is not possible owing to the principle of measurement. The respective measurements can be performed only if the frequency counter is connected directly to the components that determine the frequency of the specimen.

Instead of using a frequency standard of high stability, it is also possible to test HF radio receivers by means of the WWV signal of the National Institute of Standards and Technology (NIST), available via the air interface directly from the antenna as a receive signal. The NIST emissions originate in Fort Collins, approximately 100 km north of

Denver, Colorado. The signals are transmitted 24 hours per day and seven days per week at the exact frequencies of 5 MHz, 10 MHz and 15 MHz, each with 10 kW RF power, and at the two frequencies 2.5 MHz and 20 MHz with 2.5 kW RF power each. The signal selected should have an optimum SIR (Section III.4.8), depending on the existing propagation conditions. Detailed information on the achievable calibration accuracy and the possible applications using the WWV signals can be found in [37]. For measurements over several days taken at the same time every day with subsequently averaging the individual results, the calibration can be made with a deviation of better than $\pm 1 \cdot 10^{-9}$.

III.16   Characteristics of the Noise Squelch

One of the properties of a limiting IF amplifier used with FM is that, after demodulation and without an RF signal, it produces a noise signal noticeable at the highest AF levels. This is extremely disturbing during listening pauses and waiting times. A squelch circuit serves to suppress any loudspeaker sound when weak or no RF input signals are present. The RF input level that has to be exceeded for the squelch circuit to open the AF signal path to the loudspeaker is called the *response threshold of the noise squelch*. In order to avoid unwanted 'flatter' due to repeated switching the sound on and off, the squelch must cut off the AF path only with a somewhat lower RF input signal, the *squelch cut-in threshold* (Fig. III.104). The level difference between the activating and deactivating is called the squelch *hysteresis*.

Receivers for voice radio or for authorities and organizations responsible for placing particular emphasis on operational safety, the noise threshold for squelch is set by the manufacturer to a fixed value. This threshold is adjustable in search receivers or reconnaissance receivers (Section II.4.2) over a wide range of RF input levels. The operator

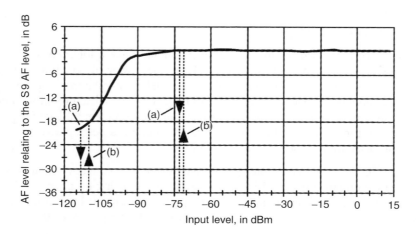

Figure III.104   Switching points of a squelch circuit over its range of variation according to the RF input level. (a) Marks the cut-in point of the squelch action and (b) represents the response threshold (the point of squelch turn-off in the AF path). The level difference between (a) and (b) is the hysteresis width.

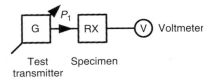

Figure III.105   Measuring arrangement for determining the characteristic of the squelch circuit.

can optimize the noise threshold for his or her particular requirements. In such units, only the hysteresis, which shifts with the adjustment across the noise threshold range, can be measured. With well-designed units it should be noted that the width of the hysteresis varies with the threshold level adjustment. In the sensitive range, in which noise and stochastic level variations are expected, the width of the hysteresis can be 3 dB, while it can change to about 1.2 dB in the upper range, in which strong and stable RF input signals must be suppressed.

Squelch circuits in F3E receiving paths react to signal frequency deviations. Sophisticated squelch circuits perform a combined evaluation of both the broad noise signal prior to near selection in a second IF stage *and* the carrier of the useful signal.

III.16.1   Measuring the Squelch Threshold

The measuring setup for determining the squelch response threshold and the cut-in point of the noise squelch is shown in Figure III.105.

Measuring procedure:

1. Tune the receiver to the frequency range to be tested.
2. Tune the test transmitter to a frequency in the centre of the IF passband.
3. Supply an RF level P_1. Slowly increase P_1 starting with the smallest possible level until the AF level jumps up or a sound suddenly becomes audible from the loudspeaker.
4. Note the level of P_1.
5. Slowly decrease the RF level P_1 until the AF level decreases rapidly or the loudspeaker is silent.
6. Note the level of P_1.

The first value determined corresponds to the squelch response threshold, and the second value represents the squelch cut-in point. The difference between the two levels is the squelch hysteresis, which can vary over the range of adjustment. The measurement should probably be repeated with a different setting or with the most sensitive and the least sensitive threshold values.

III.17   Receiver Stray Radiation

Every oscillator acts as a transmitter (of low power). Without any preventive circuitry (sufficient filtering, correct grounding, clever cable routing) and without proper screening, parts of the radio receiver can generate signals for radiation from the device. The part of the signal transmitted via the antenna socket is called the *receiver stray radiation* or oscillator radiation (sometimes with the addition 'into 50 Ω termination'). For example, one contributing factor can be insufficient LO/RF isolation of the (first) mixer. (This describes the attenuation with which the LO injection signals appears at the RF port of the mixer (Fig. III.37).) If a low-noise amplifier is configured in front of the first mixer, its backward isolation suppresses part of the returning signal. Additional attenuation is achieved by the RF selection. This is the case with an image frequency low-pass filter in heterodyne receivers with a high first IF (Section I.2.2). A three-port circulator incorporated in the signal path upstream of the (first) mixer provides additional improvement [38]. It conducts signals back to the third port, from where they are absorbed in the termination resistor to ground so that they no longer reach the antenna socket. Different circuit designs with circulators to minimize the radiation of the LO injection signal are now common in direct mixing receivers (Section I.2.3).

The spectral components of the emitted stray signals can be outside the receive frequency range of the radio receiver under test. These originate from the concept of frequency generation employed and possibly from its harmonics or from the internal mixing (Section V.4) of such signals.

National and international standards and guidelines stipulate the permissible limit values. Such stipulations often require a maximum permissible receiver stray radiation of

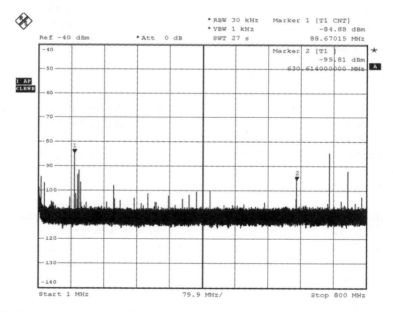

Figure III.106 Receiver stray radiation of a VLF/HF radio receiver with a receive frequency range of up to 30 MHz. (The signal levels around marker 1 are caused by unwanted incoming emission from broadcast radio on the VHF band. The measurements were performed in a normal laboratory environment with no special sheilding measures.)

−57 dBm [39]. (The pertinent EMC standards must be observed, independently of the receiver stray radiation. These also include the procedures for correctly determining those limit values.)

III.17.1   Measuring the Receiver Stray Radiation

A spectrum analyzer is connected via a short coaxial cable with the highest possible screening attenuation figure directly to the RX input in order to determine the receiver stray radiation. The spectrum analyzer should be as sensitive as possible (for small signals), with the input attenuator set to the lowest attenuation. Narrow resolution bandwidths should be used with the spectrum analyzer in order to enable a high sensitivity with sufficient separation and identification of discrete frequencies of the stray radiation. A spectrogram of the stray signals should then be taken across the frequency range to be tested. The procedure must be repeated under all relevant operating conditions to guarantee that all possibilities have been taken into account, especially with the RF preamplifier turned on/off (if applicable). The strongest level found in all measuring series represents the absolute receiver stray radiation. Ideally, such measurements should be performed in a shielded room, in which no electromagnetic radiation exists.

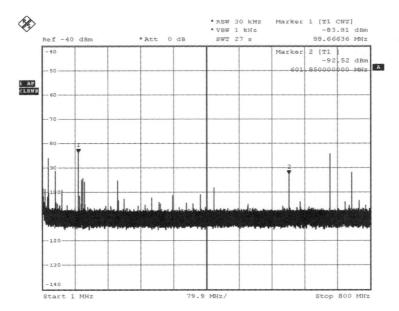

Figure III.107   Compared to Figure III.106 the receive frequency of the radio receiver tested was varied under otherwise identical conditions. Owing to the low signal level detected, the receiver shows a sufficiently low receiver stray radiation. The interfering signal frequencies different from those in Figure III.106 are undoubtedly the result of receiver stray radiation. The other interfering signal levels can result from strong transmitters in the vicinity (radiating into the test setting). The origin of the various spectral components can be reliably determined by turning the specimen off and on.

III.17.2   Measuring Problems

Due to the low signal levels to be determined and the sometimes insufficient screening of the specimen with such measurements the emission from transmitters in the vicinity (Figs. III.106 and III.107) can cause significant problems. When testing equipment with low receiver stray radiation in the absence of room shielding and without screening the measuring setup, it is very difficult to reliably identify signal levels that are definitely due to receiver stray radiation. In such cases, the receiver stray radiation can be differentiated from other signals only by turning the specimen off. (For tests covering a wide frequency range this can be very time-consuming.)

III.18   (Relative) Receive Signal Strength and S Units

Radio receivers are often provided with an indicator to display the so-called *relative receive signal strength* (commonly called relative field strength). Different units are used in such displays and range from voltage in μV and dBμV to dBm. Another evaluation criterion for the relative receive signal strength is the S unit (Tables III.6 and III.7). Initially, this was used in amateur radio services, but is now used in other radio services as well. All display modes relate to the signal strength available at the antenna input of the radio receiver.

The voltage for the display of the relative receive signal strength is often derived from the AGC control voltage (Section III.14). The higher the input signal, the lower is the required amplification at the IF level. This can be regarded as a control loop from which the signal

Table III.6   S units and the equivalent signal levels for frequencies *below* 30 MHz

S unit	Voltage at 50 Ω (μV)	Voltage level at 50 Ω (dBμV)	Power level (dBm)	Power
S1	0.20	−14	−121	794 aW
S2	0.39	−8	−115	3.16 fW
S3	0.79	−2	−109	12.6 fW
S4	1.58	4	−103	50 fW
S5	3.13	10	−97	200 fW
S6	6.25	16	−91	794 fW
S7	12.50	22	−85	3.16 pW
S8	25	28	−79	12.5 pW
S9	50	34	−73	50 pW
S9 + 10 dB	158	44	−63	500 pW
S9 + 20 dB	500	54	−53	5 nW
S9 + 30 dB	1,583	64	−43	50 nW
S9 + 40 dB	5,000	74	−33	500 nW
S9 + 50 dB	15,830	84	−23	5 μW
S9 + 60 dB	50,000	94	−13	50 μW
S9 + 70 dB	158,300	104	−3	500 μW

driving the meter (the S meter, for example) can be taken. Due to the complicated circuitry for supplying the control voltage required to ensure that the indicated value corresponds to the real receive signal level, this method has its limits. Another difficulty is the often significantly lower dynamics of the control voltage compared with the approximately 110 dB of the S meter dynamics (S1 to S9 + 60 dB) [40]. Since no AGC is required for processing class F3E emissions, there is generally no parameter available to provide a value proportional to the signal strength in a simple way (see also Section III.14). Professional solutions use integrated circuits with logarithmic level detectors. In super-het receivers (Section I.2.1) the uncontrolled signal fed to the IF stage is used for this purpose.

Apart from a few notable exceptions, most of the low-priced and semi-professional radio receivers show a similar behaviour regarding the indication of the receive signal strength (Fig. III.108). While a clear insensitivity can be detected in the lower range (the input level must be significantly higher than required for an adequate S unit), any value above S7 indicates suitable conformity with the respective levels. Manufacturers nearly always calibrate their equipment for S9, so that this point can be taken as a reliable indication (Fig. III.109). If an RF preamplifier that may be built in or the RF attenuator can be separately activated, this also affects the displayed value sometimes and produces indication errors. The resulting change in the reading is simply false and is caused by an inadequate circuit design.

It is common practice to use this value for the estimation of the strength of incident signals. *In fact, such estimations have a limited information value and are mainly suitable only for relative comparisons.*

Table III.7   S units and the equivalent signal levels for frequencies *above* 30 MHz

S unit	Voltage at 50 Ω (µV)	Voltage level at 50 Ω (dBµV)	Power level (dBm)	Power
S1	0.020	−34	−141	7.94 aW
S2	0.039	−28	−135	31.6 aW
S3	0.079	−22	−129	126 aW
S4	0.158	−16	−123	500 aW
S5	0.313	−10	−117	2 fW
S6	0.625	−4	−111	7.94 fW
S7	1.250	2	−105	31.6 fW
S8	2.500	8	−99	126 fW
S9	5	14	−93	500 fW
S9 + 10 dB	15.8	24	−83	5 pW
S9 + 20 dB	50	34	−73	50 pW
S9 + 30 dB	158	44	−63	500 pW
S9 + 40 dB	500	54	−53	5 nW
S9 + 50 dB	1,583	64	−43	50 nW
S9 + 60 dB	5,000	74	−33	500 nW
S9 + 70 dB	15,830	84	−23	5 µW

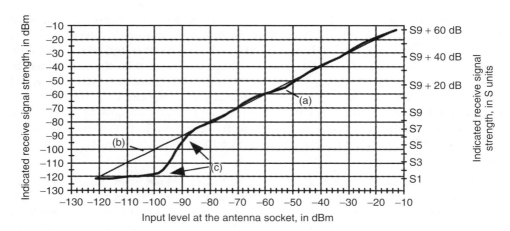

Figure III.108 Diagram indicating the relative signal strength with the voltage for the indicator taken from the AGC. The curve of the S units exemplifies how the S units can be used with all-wave receivers (Section II.3.4) for indicating the signal strength. The curves show (a) the typical and (b) the ideal progression, while (c) marks the clearly insufficiently sensitive lower range. With class F3E emissions this insensitive range is even more distinct if the circuit is not adapted accordingly.

Only the K factor (also called antenna factor or transducer figure) provides a means to determine the field strength induced by an outstation in the space around the receiving antenna from the relative value indicated on the radio receiver. For an antenna of known antenna gain the K factor can be calculated in 50 Ω systems [41]:

$$K(f_{op}) = -29.8 \text{ dB}(1/\text{m}) + 20 \cdot \lg \left(\frac{f_{op}}{10^6 \text{ Hz}} \right) - G_{dBi \text{ ant}} \qquad (III.42)$$

where
$K(f_{op})$ = transducer figure of the antenna in use, depending on the actual operating
frequency (f_{op}), in dB(1/m)
f_{op} = operating frequency, in Hz
$G_{dBi \text{ ant}}$ = antenna gain figure of the antenna in use, in dBi

(In 60 Ω systems it is -30.6 dB(1/m) instead of -29.8 dB(1/m) and in 75 Ω systems it is -31.5 dB(1/m).) The transducer figure thus depends on the frequency and can be determined from the operating frequency through the antenna gain, which is also frequency-dependent. Losses occurring over the antenna cable must be added to the transducer figure. The true field strength in space around the receiving antenna is then given by

$$E = K(f_{op}) + V_{S \text{ unit}} \qquad (III.43)$$

where
E = field strength in the space around the antenna in use, in dB(μV/m)
$K(f_{op})$ = transducer figure of the antenna in use, depending on the actual operating
frequency (f_{op}), in dB(1/m)
$V_{S \text{ unit}}$ = signal strength indicated on the radio receiver, in dBμV

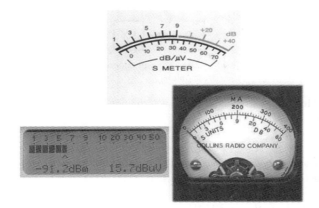

Figure III.109 Display scales for indicating the relative receive signal strength as used in typical all-wave receivers (Sections II.3.2 and II.3.4) and receiving paths in radio telecommunication systems (with frequency ranges below 30 MHz).

For a receive frequency of 430.2 MHz, the calibrated S meter of the receiver with 50 Ω input indicates the value S8.5 when using a multi-element directional antenna oriented to the outstation. (The antenna is assumed to have a 9 dBi antenna gain figure and a 4 dB attenuation figure due to the antenna feeder cable, while the other losses from coaxial connector transitions are assumed to be 0.3 dB in total.) The transducer figure of the antenna system is

$$K(f_{\mathrm{op}} = 430.2\ \mathrm{MHz}) = -29.8\ \mathrm{dB}(1/\mathrm{m}) + 20 \cdot \lg\left(\frac{430.2\ \mathrm{MHz}}{1\ \mathrm{MHz}}\right) - 9\ \mathrm{dBi}$$

$$= 13.9\ \mathrm{dB}(1/\mathrm{m})$$

The voltage level of 8 dBμV corresponding to S8 can be taken from Table III.7. The value for an additional half S step is 3 dB higher and is equal to 11 dBμV. The field strength originating from the transmitting station and measured at the point of reception is

$$E = 13.9\ \mathrm{dB}(1/\mathrm{m}) + 4\ \mathrm{dB} + 0.3\ \mathrm{dB} + 11\ \mathrm{dB\mu V} = 29.2\ \mathrm{dB}(\mu V/\mathrm{m})$$

or 28.8 μV/m expressed as a voltage.

III.18.1 Definitions and Predetermined Levels of S Units

The reference level of S9 is predefined as S9 = 50 μV at 50 Ω below 30 MHz and S9 = 5 μV at 50 Ω above 30 MHz. Each S unit differs by 6 dB from the next value above and below. Higher signal strengths than S9 relevant for higher input values are given directly in 10 dB increments above the 'reference level S9' and no longer in S units. For each S unit there is an equivalent voltage at the antenna socket.

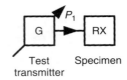

Figure III.110 Measuring arrangement for determining the accuracy of the relative signal strength indication.

A distinction is made only on the basis of the frequency ranges below 30 MHz and above 30 MHz. The reasons for this are as follows:

(a) The higher path loss above 30 MHz (Fig. II.1) prevents signals as strong as in the medium and short wave ranges.
(b) The received noise intensity due to external noise (Section III.4.10) at higher receive frequencies is weaker, and smaller signals can be detected, which provides another evaluation criterion.

The relation between the S units and the corresponding levels at the receiver input is shown in Tables III.6 and III.7.

III.18.2   Measuring the Accuracy of the Relative Signal Strength Indication

The measuring setup for determining the deviations in instrument readings of the relative receive signal strength is shown in Figure III.110.

Measuring procedure:

1. Tune the receiver to the frequency range to be tested.
2. Tune the test transmitter to a frequency in the centre of the IF passband.
3. Supply an RF level P_1 corresponding to the indication set-point. Gradually increase P_1.
4. Always note the actual value and the set-point value.

The difference between the actual and the set-point values is the deviation from the correct indication (inaccuracy). For a graphical evaluation it may be useful to enter the values determined in a diagram (similar to Fig. III.108).

III.18.3   Measuring Problems

The accuracy achievable for determining indicating errors is limited by level variations of the test transmitter used.

Owing to the mismatch of the test transmitter output (output matching) and the mismatch of the receiver input (Section III.3) there is a measuring uncertainty. The uncertainty due to mismatching can be expressed in percent as

$$M_\% = 100\% \cdot \left(\left(1 \pm \frac{SWR_{TTX} - 1}{SWR_{TTX} + 1} \cdot \frac{SWR_{RX} - 1}{SWR_{RX} + 1} \right)^2 - 1 \right) \qquad \text{(III.44)}$$

where
$M_\%$ = measuring uncertainty due to mismatching, in %
SWR_{TTX} = standing wave ratio of the test transmitter output, dimensionless
SWR_{RX} = standing wave ratio of the receiver input, dimensionless

From this, the level measuring uncertainty (level error) can be calculated directly and is given by the expression

$$M_{dB} = 20 \cdot \lg \sqrt{\frac{M_\%}{100\%} + 1} = 20 \cdot \lg \left(1 \pm \frac{SWR_{TTX} - 1}{SWR_{TTX} + 1} \cdot \frac{SWR_{RX} - 1}{SWR_{RX} + 1} \right) \qquad \text{(III.45)}$$

where
M_{dB} = level measuring uncertainty due to mismatching, in dB
SWR_{TTX} = standing wave ratio of the test transmitter output, dimensionless
SWR_{RX} = standing wave ratio of the receiver input, dimensionless

In the worst case, the sum of the maximum level deviation of the test transmitter used and the level error due to mismatching is the overall error. The deviations of the instruments indicating the relative receiver signal strength cannot be determined with a higher accuracy.

The receiver input matching of the specimen in the useful frequency band is specified for example by using a standing wave ratio of 2.5. According to its data sheet, the test transmitter used for determining the indication accuracy has an output matched to its test frequency of SWR = 1.3 and a possible level tolerance of ±0.45 dB. When reading the relative receive field strength on the receiver under test, the measuring uncertainty due to mismatching is

$$M_\% = 100\% \cdot \left(\left(1 + \frac{1.3 - 1}{1.3 + 1} \cdot \frac{2.5 - 1}{2.5 + 1} \right)^2 - 1 \right) = +11.5\%$$

or

$$M_\% = 100\% \cdot \left(\left(1 - \frac{1.3 - 1}{1.3 + 1} \cdot \frac{2.5 - 1}{2.5 + 1} \right)^2 - 1 \right) = -10.9\%$$

or, expressed as the level error

$$M_{dB} = 20 \cdot \lg \sqrt{\frac{11.5\%}{100\%} + 1} = 20 \cdot \lg \left(1 + \frac{1.3 - 1}{1.3 + 1} \cdot \frac{2.5 - 1}{2.5 + 1} \right) = +0.47 \text{ dB}$$

or

$$M_{\mathrm{dB}} = 20 \cdot \lg \sqrt{\frac{-10.9\%}{100\%} + 1} = 20 \cdot \lg \left(1 - \frac{1.3 - 1}{1.3 + 1} \cdot \frac{2.5 - 1}{2.5 + 1} \right) = -0.5 \, \mathrm{dB}$$

In this example for the determination of the indicator accuracy a possible overall error of 0.5 dB + 0.45 dB = 0.95 dB remains.

III.19 AM Suppression in the F3E Receiving Path

The information content of an F3E modulated signal is contained entirely in its frequency variations. One of the main advantages of this class of emission is its high resistance to amplitude changes. This makes it comparatively robust against amplitude interferences occurring along the transmission path (like fading or interferences in mobile operation). This advantageous property is achieved in demodulation with a wide-band (amplitude) limiter of very short settling time. With high amplitude variations in the RF input signal, the AF output signal still contains signal components with the frequency of the amplitude variation. This serves to reduce the signal-to-interference ratio. The lower the disturbance caused, the higher is the *AM suppression*. In this respect, the commercially available radio receivers or receiving modules differ widely.

An investigation by measurement is done by feeding an F3E signal with nominal modulation (Section III.2) to the receiver, while this F3E signal carries an additional amplitude-modulated tone frequency of a certain modulation depth. The test transmitter used must be capable of simultaneous AM and FM modulation (Figs. III.111 and III.112).

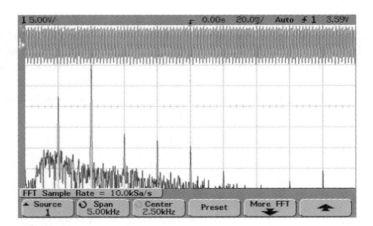

Figure III.111 AF output signal of an F3E receiver with low AM suppression. The separation between the 1 kHz useful signal and the other 500 Hz signal resulting from insufficient AM suppression is only 14 dB. The harmonics appearing to the right of the useful signal arise from the demodulation (compare with Figs. III.31 and III.32). (Y axis: 10 dB/div., X axis: 500 Hz/div.; the upper part shows the demodulated signal in the time domain.)

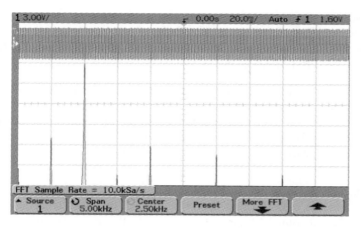

Figure III.112   AF output signal of an F3E receiver with strong AM suppression. The separation between the 1 kHz useful signal and the other 500 Hz signal occurring is 35 dB in this case. Similar to Figure III.111, the useful signal supplied to the F3E receiver carries an additional 500 Hz interfering modulation with a modulation depth of 30%. (Y axis: 10 dB/div., X axis: 500 Hz/div.; the upper part shows the demodulated signal in the time domain.)

III.19.1   Measuring the AM Suppression

The measuring setup for determining the AM suppression of a F3E receiving path is shown in Figure III.113.

Measuring procedure:

1. Tune the receiver to the frequency range to be tested.
2. Connect a selective level meter (via a weighting filter (Section III.4.8) if necessary) to the AF output of the receiver.
3. Tune the test transmitter to the receive frequency and, by doing so, feed a defined RF level P_1 modulated with nominal modulation to the receiver. (A P_1 of -107 dBm is often used. 60 dBμV is a common value in VHF FM broadcasting technology.)

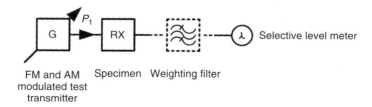

Figure III.113   Measuring arrangement for determining the AM suppression in F3E receiving paths.

4. Add an additional AM modulation with a defined frequency and defined modulation depth to the RF level P_1. (A modulation depth of 30% and a modulation frequency of 400 Hz or 500 Hz are often used. Modulations with 400 Hz FM and 1,000 Hz AM are common in VHF FM broadcasting technology.)

The level difference between the two demodulated and separately measured tone frequencies represents the AM suppression. Instead of using the selective level meter or voltmeter for true effective values with an upstream bandpass filter for isolating the frequencies to be measured, FFT analysis can also be performed. The FFT analysis can be performed with a modern digital storage oscilloscope (Fig. III.111) or with specific software in combination with a PC soundcard.

If this equipment is not available, a possible alternative is to increase the modulation depth in step 4 until the SINAD (Section III.4.8) at the AF output is reduced to 20 dB. This does not provide a measure of the AM suppression in the true sense, but does represent a parameter for the (subjective) comparison after a suitable measuring series with various different units is performed.

III.20 Scanning Speed in Search Mode

In automatic search reception (by search runs) for the quick detection of occupation and for monitoring signal activities within predetermined frequency ranges, a sequential search (Section II.4.2) is used to detect signals at unknown transmit frequencies within time spans as short as possible. The following sections therefore concentrate on single-channel receiver systems with narrow-band receive channels. The receive centre frequency is tuned gradually across the frequency range monitored, and the occupation and incident signal intensity of the discrete channels tested are determined by measuring the energy in the receive bandwidth (Section III.6.1).

The achievable scanning speed (or sweep speed) depends on the search objective. It is therefore a compromise between the width of the frequency range to be monitored or searched, the tuning increment or selected channel pattern (frequency resolution) and the receive bandwidth. The scanning speed of a radio receiver or scanner receiver (Section II.4.2) is determined by the tuning time of the preselector, the settling time of the narrowest selection filter in the receiving path (according to Equation (II.1)), and by the processing and response/control times of digital processing. In order to achieve particularly high search or scanning speeds, the frequency range under observation must be limited to the bandwidth of the preselector or the preselection must be bypassed. A correspondingly short settling or transient time of the oscillators in the receiver is essential. This can be achieved by optimizing the receiver design using, for example, two or three tuning oscillators with fast switch-over capability. When the receive frequency is generated by one oscillator, the signals for the following tuning step (or even the signal for the step after this) can be generated simultaneously and have time to settle, so that these are available for tapping without delay.

The scanning speed is an important receiver parameter used in scan mode (Section II.4.2), since it determines the uncertainty in detecting an unknown short-term emission. For a given scanning period the scanning speed determines the number of discretely verifiable frequency channels and the time gaps during signal acquisition. According to [42] the specifications for such receivers should be with reference to a detection probability (Table II.5) of 95%. Errors of 5 dB in the relative receive signal strength (Section III.18) in occupied channels and a frequency tolerance (Section III.15) extended by the receive bandwidth in Hz, both relative to the static receiving mode, are tolerable. Correct specifications should state the speed in MHz and be related to the widest observable frequency range.

III.20.1   Measuring the Scanning Speed

The measuring setup for determining the scanning speed is shown in Figure III.114.

Measuring procedure:

1. Select the start and stop frequency of the frequency range to be monitored and other scan parameters (tuning increments, receive bandwidth) at the receiver. (According to [42], for receive frequency ranges below 30 MHz a receive bandwidth of 5 kHz and above 30 MHz a receive bandwidth of 25 kHz shall be used. If only another receive bandwidth can be selected on the equipment under test, use the next narrower filter setting.)
2. Tune a test transmitter with burst mode and external triggering capability to a discrete frequency channel within the scanned frequency range. Prepare the test transmitter for emitting an RF burst level P_1 18 dB above the operational sensitivity at 12 dB SINAD (Section III.4.8) of the specimen.
3. Adjust the test transmitter so as to emit a single RF burst released by a trigger pulse.

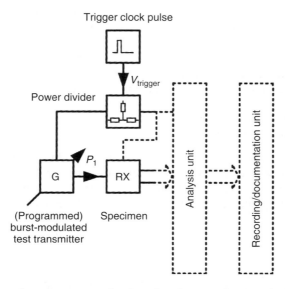

Figure III.114   Measuring arrangement for detecting the scanning speed of receivers in search mode.

4. Scan the frequency range to be monitored with the specimen once and record the single RF burst.
5. Repeat steps 3 and 4 and shorten the duration of the RF burst signal cyclically until the RF burst is just barely detected.
6. Repeat steps 3 and 4 over several cycles to verify the (specified) detection probability (95% is a common value).

For the given parameters, the scanning speed of the specimen can be estimated on the basis of the minimum RF burst length as determined in step 5:

$$v_{\text{scan}} = \frac{f_{\text{stop}} - f_{\text{start}}}{t_{\text{burst}}} \qquad\qquad (\text{III.46})$$

where
v_{scan} = scanning speed, in MHz/s
f_{stop} = stop frequency of the frequency range to be monitored, in MHz
f_{start} = start frequency of the frequency range to be monitored, in MHz
t_{burst} = length of the RF burst, in s

If the frequency range between 117.975 MHz and 144 MHz that is primarily assigned to air traffic radio is scanned for signal activity and short-term signals down to 28 ms are reliably detected, the receiver under test uses a scanning speed of

$$v_{\text{scan}} = \frac{144\,\text{MHz} - 117.975\,\text{MHz}}{0.028\,\text{s}} = 929\,\text{MHz/s}$$

Extended measuring procedure:

7. Prepare the programmable test transmitter with burst mode and external triggering capability for emission in the frequency range scanned with more than 50 RF bursts of the duration determined in step 5.
8. Scan the frequency range to be monitored with the specimen and record the RF bursts released by the trigger pulses in order to verify the (specified) detecting probability in the dynamic receiving mode (95% is a common value).

Perform an additional test in which, unlike the previous test, all channels of the frequency ranged monitored are occupied.

Due to the high speed, this method requires automatic evaluation of the results supplied from the receiver (event count, detected centre frequency and level of the RF bursts). In modern search receivers such evaluations are part of the equipment-based analysis software. Alternatively, the data can/must be made available via an interface for external evaluation.

References

[1] Michael Gabis, Ralf Rudersdorfer: Aktuelle digitale Funkstandards im transparenten Vergleich zum analogen FM-Sprechfunk – Teil 1 bis Teil 2 (Current Digital Radio Standards Transparently Compared with Analog FM Voice Radio – Part 1 and Part 2); UKWberichte 4/2007, pp. 195–208, UKWberichte 2/2008, pp. 107–119; ISSN 0177–7513

[2] Claude E. Shannon, Warren Weaver: The Mathematical Theory of Communication; 1st edition; University of Illinois Press 1949; ISBN 0-252-72546-8

[3] Klaus von der Heide: Digitale Kommunikation – Grundlagen und Grenzen (Digital Communications – Basic Principles and Limitations); CQ DL Spezial 9/2011, pp. 4–10; ISSN 0178-269X

[4] Ian A. Glover, Peter M. Grant: Digital Communications; 1st edition; Prentice Hall 1998; ISBN 0-13-565391-6

[5] Wolf-Jürgen Becker, Karl W. Bonfig, Klaus Höing, editors: Handbuch elektrische Messtechnik (Handbook for Electrical Measurement Engineering); 1st edition; Hüthig Verlag 1998; ISBN 3-7785-2740-1

[6] Hans H. Meinke, Friedrich-Wilhelm Gundlach, editors: Taschenbuch der Hochfrequenztechnik (Handbook for Radio Frequency Technology); 5th edition; Springer Verlag 1992; ISBN 3-540-54717-7

[7] Friedrich Landstorfer, Heinrich Graf: Rauschprobleme in der Nachrichtentechnik (Noise Problems in Communications Technology); 1st edition; Oldenbourg Verlag 1981; ISBN 3-486-24681-x

[8] Ralf Rudersdorfer: Messung kritischer Spezifikationen an Signalquellen – Teil 1 und Teil 2 (Measuring Critical Specifications at Signal Sources – Part 1 and Part 2); funk 5/2005, pp. 58–63, funk 6/2005, pp. 60–64; ISSN 0342-1651

[9] Frank Sichla, Ralf Rudersdorfer: So misst die 'funk' – Meßvorschrift für KW-Empfänger, -Sender, -Transceiver-Empfangs-und Sendeteile (The Way 'funk' Measures – Measuring Instructions for Receiving and Transmitting Components of SW Receivers, Transmitters and Transceivers); funk 2/2002, pp. 68–71; ISSN 0342-1651

[10] Frank Sichla: Im Größenrausch der Rauschgrößen (Much Noise about Noise Issues); funk 12/2001, pp. 59–63; ISSN 0342-1651

[11] Ralf Rudersdorfer: Empfängercharakteristika und Großsignalverhalten analoger Empfänger (Receiver Characteristics and Large-Signal Behavior of Analog Receivers); qsp – Organ des Österreichischen Versuchssenderverbandes 7/2001 –26/7, pp. 54–63

[12] International Telecommunication Union (ITU), publisher: Psophometer for use on telephone-type circuits; ITU Recommendation O.41/p. 53 10/1994

[13] Erich Stadler: Modulationsverfahren (Modulation Methods); 8th edition; Vogel Buchverlag 2000; ISBN 3-8023-1840-4

[14] Ulrich Graf, Hans-Hellmuth Cuno: Messung von FM-Geräten (Measurement of FM Equipment); CQ DL 7/2000, pp. 499–502; ISSN 0178-269X

[15] Norbert Ephan, Anton Ilsanker, Bernhard Liesenkötter: Rauschen von Satellitenempfangsantennen – eine Faustregel für Rauschkalkulationen (Noise Level of Satellite Receiving Antennas – A Rule of Thumb for Noise Calculations); Rundfunktechnische Mitteilungen 6/1989 –33, pp. 292–296; ISSN 0035–9890

[16] Hans D. Baehr, Karl Stephan: Wärme- und Stoffübertragung (Heat Transfer and Material Transmission); 4th edition; Springer Verlag 2004; ISBN 3-540-40130-X

[17] International Telecommunication Union (ITU), publisher: Radio noise; ITU Recommendation P.372 4/2003

[18] Peter A. Bradley: Whither noise and interference?; Proceedings of the Second IEE Colloquium on Frequency Selection and Management Techniques for HF Communications London 1999, pp. 6.1–6.8

[19] Ulf Schneider: Grenzempfindliche passive Breitbandempfangsantenne 0,1 bis 30 MHz, als Sendeantenne nutzbar von 14 bis 30 MHz (Boundary-Sensitive Passive Broadband Receiving Antenna 0.1 to 30 MHz, Usable as a Transmission Antenna from 14 to 30 MHz); manuscript 2009, pp. 1–6

[20] Jochen Jirmann: Aktive elektrische/magnetische Empfangsantennen mit gängigen Transistoren (Active Electric/Magnetic Receiving Antennas with Common Transistors); manuscripts of speeches from the AFTM Munich 2010, pp. 12.1–12.10

[21] Christoph Rauscher: Grundlagen der Spektrumanalyse (Fundamentals of Spectrum Analysis); 2nd edition; Rohde&Schwarz in-house publishers 2004; PW 0002.6629.00

[22] Ulrich Freyer: Messtechnik in der Nachrichtentechnik (Measurement Technology in Communications Engineering); 1st edition; Carl Hanser Verlag 1983; ISBN 3-446-13703-3

[23] Acterna/Schlumberger ENERTEC, publisher: Handbuch Stabilock 535A SSB-Messplatz (Handbook Stabilock 535A SSB Test Set); 1985 edition, Part 4 Applications

[24] Gerhard Krucker: Elektronische Signalverarbeitung (Electronic Signal Processing); manuscript of the FH Bern 2000, pp. 1–179

[25] Wolf D. Schleifer: Hochfrequenz-und Mikrowellen-Meßtechnik in der Praxis (The Practical Approach to Radio Frequency and Microwave Measuring Technology); 1st edition; Dr. Alfred Hüthig Verlag 1981; ISBN 3-7785-0675-7

[26] Ulrich Graf: Empfänger-Intermodulation – Teil 1 bis Teil 3 (Receiver Intermodulation – Part 1 to Part 3); CQ DL 6/2002, pp. 436–438, CQ DL 7/2002, pp. 504–507, CQ DL 8/2002, pp. 588–591; ISSN 0178-269X

[27] Henning C. Weddig: Messungen des Intermodulationsverhaltens von Quarzfiltern (Measuring the Intermodulation Behaviour of Quartz Filters); manuscript of speeches at the VHF Convention, Weinheim 2003, pp. 23.1–23.11

[28] Detlef Lechner: Kurzwellenempfänger (Short-Wave Receivers); 2nd edition; Militärverlag der Deutschen Demokratischen Republik 1985

[29] Hans Zahnd: Bestimmung des Intermodulations – Verhaltens (Determining the Intermodulation Behavior); adat Engineering Note EN01 7/2002

[30] International Telecommunication Union (ITU), publisher: Test procedure for measuring the 3rd order intercept point (IP3) level of radio monitoring receivers; ITU Recommendation SM.1837 12/2007

[31] Kurt Hoffelner: RX-Messtechnik (RX Measurement Technology); manuscript of speech at the OAFT Neuhofen/Ybbs 2008, pp. 1–19

[32] Wes Hayward: Receiver Dynamic Range; QST 7/1975, pp. 15–22; ISSN 0033–4812

[33] Robert E. Watson: Receiver Dynamic Range – Part 1 and Part 2; Watkins-Johnson Tech-note 1/1987 – Vol. 14, Watkins-Johnson Tech-note 2/1987 – Vol. 14

[34] Otto Zinke, Heinrich Brunswig, editors: Hochfrequenztechnik 2 (Radio Frequency Technology 2); 4th edition; Springer Verlag 1993; ISBN 3-540-55084-4

[35] Reiner S. Thomä, editor: Messung von Empfängerkenngrößen (Measuring the Receiver Characteristics); manuscript for the information electronics training of the TU Illmenau 5/2000, pp. 1–15

[36] Heinz Lindenmeier, Jochen Hopf: Kurzwellenantennen (Short-Wave Antennas); 1st edition; Hüthig Verlag 1992; ISBN 3-7785-1996-4

[37] Glenn K. Nelson, Michael A. Lombardi, Dean T. Okayama: NIST Time and Frequency Radio Stations: WWV, WWVH, and WWVB; National Institute of Standards and Technology Special Publication 250–67 2005

[38] Ulrich Tietze, Christoph Schenk: Halbleiter-Schaltungstechnik (Circuit Designs Using Semi-Conductors); 13th edition; Springer Verlag 2010; ISBN 978-3-642-01621-9

[39] Volker Jung: Handbuch der Telekommunikation (Handbook for Telecommunications); 1st edition; Springer Verlag 2002; ISBN 3-540-42795-3

[40] Ralf Rudersdorfer: Der S-Wert in Theorie und Praxis (The S Value in Theory and Practice); funk 11/2002, pp. 61–62; ISSN 0342–1651

[41] Karl H. Hille, Alois Krischke: Das Antennen-Lexikon (The Antenna Dictionary); 1st edition; Verlag für Technik und Handwerk 1988; ISBN 3-88180-304-1

[42] International Telecommunication Union (ITU), publisher: Test procedure for measuring the scanning speed of radio monitoring receivers; ITU Recommendation SM.1839 12/2007

Further Reading

German Institute for Standardization (DIN), publisher: Einheiten – Teil 1 Einheitennamen, Einheitenzeichen (Measuring Units – Part 1 Unit Names, Unit Symbols); DIN standard DIN 1301–1 10/2002

German Institute for Standardization (DIN), publisher: Logarithmische Größen und Einheiten –Teil 2 Logarithmierte Größenverhältnisse, Maße, Pegel in Neper und Dezibel (Logarithmic Quantities and Units – Part 2 Logarithmic Ratios, Measures, Levels in Nepers and Decibels); DIN Norm DIN 5493-2 9/1994

Karl-Heinz Gonschorek: EMV für Geräteentwickler und Systemintegratoren (EMC for Equipment Developers and System Integrators); 1st edition; Springer Verlag 2005; ISBN 978-3-540-23436-4

Rudolf Grabau: Funküberwachung und Elektronische Kampfführung – suchen, aufnehmen, peilen, stören, schützen (Radio Surveillance and Electronic Warfare – Searching, Recording, Direction Finding, Interfering, Protecting); 1st edition; franckh Verlag 1986; ISBN 3-440-05667-8

Ulrich Graf, Hans-Hellmuth Cuno: Warum so messen? (Why this Measurement Technique?); CQ DL 11/1998, pp. 861–863; ISSN 0178-269X

Ulrich Graf: Performance Specifications for Amateur Receivers of the Future; QEX 5+6/1999, pp. 43–49; ISSN 0886–8093

Ulrich Graf: Was sollen 'gute' Amateurempfänger können? – Teil 1 bis Teil 4 (What is Expected of a 'Good' Amateur Receiver? – Part 1 to Part 4); CQ DL 12/1997, pp. 940–943, CQ DL 1/1998, pp. 36–38, CQ DL 2/1998, pp. 122–124, CQ DL 3/1998, pp 227; ISSN 0178-269X

International Telecommunication Union (ITU), publisher: Parameters of and measurement procedures on H/V/UHF monitoring receivers and stations; ITU Report SM.2125 1/2008

International Telecommunication Union (ITU), publisher: Test procedure for measuring the noise figure of radio monitoring receivers; ITU Recommendation SM.1838 12/2007

International Telecommunication Union (ITU), publisher: Test procedure for measuring the properties of the IF filter of radio monitoring receivers; ITU Recommendation SM.1836 12/2007

International Telecommunication Union (ITU), publisher: Test procedure for measuring the sensitivity of radio monitoring receivers using analogue-modulated signals; ITU Recommendation SM.1840 12/2007

International Telecommunication Union (ITU), publisher: Use of the decibel and the neper in telecommunications; ITU Recommendation V.574-4 5/2000

André Jamet: RST Code and S-Meter revisited; VHF Communications 1/2009, pp. 21–24; ISSN 0177–7505

Anthony R. Kerr, Marc J. Feldman, Shing-Kuo Pan: Receiver Noise Temperature, the Quantum Noise Limit, and the Role of the Zero-Point Fluctuations; Proceedings of the Eighth International Symposium on Space Terahertz Technology Harvard 1997, pp. 101–111

Ulrich L. Rohde: Theory of Intermodulation and Reciprocal Mixing: Practice, Definitions and Measurements in Devices and Systems – Part 1 and Part 2; QEX 11+12/2002, pp. 3–15, QEX 1+2/2003, pp. 21–31; ISSN 0886–8093

Tony J. Rouphael: RF and Digital Signal Processing for Software-Defined Radio; 1st edition; Elsevier Newnes 2008; ISBN 978-0-7506-8210-7

Ralf Rudersdorfer: Wichtige Empfängerkennwerte verständlich gemacht (Important Receiver Characteristics Explained for Easy Understanding); funk 5/2001, pp. 38–40; ISSN 0342–1651

Werner Schnorrenberg: Spektrumanalyse in Theorie und Praxis (Spectrum Analysis in Theory and Practice); 1st edition; Vogel Buchverlag 1990; ISBN 3-8023-0290-7

Manfred Thumm, Werner Wiesbeck, Stefan Kern: Hochfrequenzmesstechnik (Radio Frequency Measurement Engineering); 1st edition; B. G. Teubner Verlag 1997; ISBN 3-519-06360-3

Robert E. Watson: Use one figure of merit to compare all receivers; Microwaves & RF 1/1987 –pp. 99–108; ISSN 0745–2993

IV

Practical Evaluation of Radio Receivers (A Model)

IV.1 Factual Situation

New experiences are gained with every *practical test reception* with a (newly developed) radio receiver. In fact, receivers behave at times differently than expected from the pertinent characteristic parameters (Part III) specified for receiver operation (under restricted conditions).

While the performance data of receivers are very clearly described by the receiver characteristics regarding their large-signal behaviour (Section III.12) in exactly standardized radio services, this is not the case with receivers designed for applications in barely or poorly coordinated frequency bands or radio services.

For example, for a radio service operating with a fixed channel spacing any intermodulation (Section III.9) or blocking (Section III.8) can be expected only with interfering signals having at least the same frequency separation (or a multiple of this). The receiver characteristics of interest and specified can then actually describe the relevant receiving conditions.

Utterly different is the situation with equipment like search receivers (Section II.4.2) typically used in radio intelligence, short-wave receivers, or receivers for satellite communications in amateur radio service (the technical term is uncoordinated multiple access). The frequency separation between the useful signal and the interfering signal(s) can vary as much as the possible level differences. (The frequency range below 30 MHz is particularly complex and demanding (Fig. IV.1). The considerations described below will concentrate on this frequency range but of course apply correspondingly to other receive frequency ranges.)

IV.2 Objective Evaluation of Characteristics in Practical Operation

The facts described in Section IV.1 prompted the development of procedures to comprehensively and clearly evaluate the performance data of receivers (for these specific

Radio Receiver Technology: Principles, Architectures and Applications, First Edition. Ralf Rudersdorfer.
© 2014 Ralf Rudersdorfer. Published 2014 by John Wiley & Sons, Ltd.

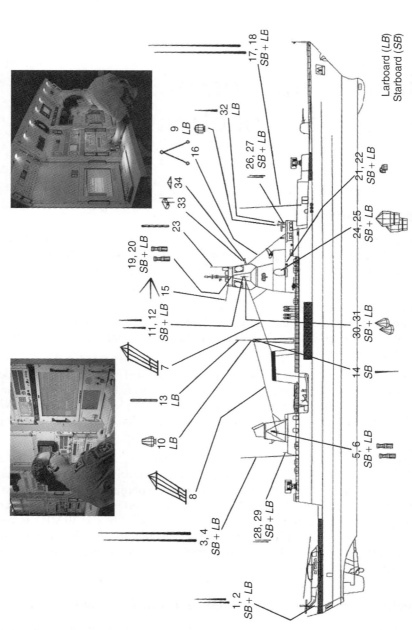

Figure IV.1 Exploded view of telecommunication equipment and receiving antennas onboard a marine vessel [1]. Owing to the tight space, the various frequencies used for transmission and reception simultaneously, and the high transmit power, the receivers have to cope with very high receive voltages. Since weaker useful signals must also be demodulated intelligibly from the incoming sum signal mix such signal scenarios place very high demands on radio receivers. (Company photographs of Rohde&Schwarz.)

applications) in a transparent fashion. This Part describes a model for performing comparisons under practical conditions: The 'Table of operational *PRACTICE*' (see Table IV.3). The relevant testing conditions are also specified.

In view of the purpose and relevance of characteristic receiver parameters like those specified by the manufacturer or stated in test reports published in trade magazines and on the basis of the descriptions in [2] and [3] one can draw the following conclusions:

(a) With high-performance measuring equipment a *certain signal scenario* can be reproduced in the laboratory *under repeatable conditions*. For this purpose, the various (receiving) situations are reproduced one after the other by means of metrological equipment in order to scrutinize the various technical effects to obtain a more exact and more comparable characterization. The more effort invested, the more meaningful are the results.

(b) In *practical receive mode* there are some additional effects that can (could) be measured to a limited degree with considerable effort, but which may also cause situations that are not foreseeable or adequately characterized by the data specified.

(c) Particularly important is the 'interaction' (Section III.12.1) of the individual parameters determined in quality and quantity – the characteristic receiver parameters. For this particular reason the receiving practice, in other words *practical evaluation under real operating conditions*, plays a very important role.

(d) Regardless of how often one measures and graphically documents the AF frequency response, the signal must sound pleasant to the ear when audibly reproducing the demodulated information. The type of loudspeaker used, its installation position, and its sound generation capability can be highly influential on the sound quality and its subjective perception.

IV.2.1 Hardly Equal Conditions

The following remarks are relevant regarding Section IV.2(a): Propagation conditions and frequency band occupancy often change within short time intervals. An objective assessment of the listening practice alone would not be very meaningful since on days with particularly poor propagation conditions one would not be tempted to provoke a strenuous signal scenario long enough to reach the performance limits of the receiver (Fig. IV.2). Especially in this respect, contrary to listening alone, the measuring results are of utmost importance for the comparison of equipment, since this allows one to subject any unit to be tested to the same receiving situation at the antenna socket.

IV.2.2 No Approximation Possible

Regarding Section IV.2(b) from time to time, sometimes from well-known specialists, one hears comments like 'receivers cannot be fully understood by metrological methods' or 'the most important thing for me is what I can hear'. (However, it needs to be emphasized that there is no professional shortcut for the comprehensive metrological evaluation of radio receiver performance. Hearing comparisons can provide *additional* information on the typical behaviour of a receiver in accompanying tests. This is especially true for

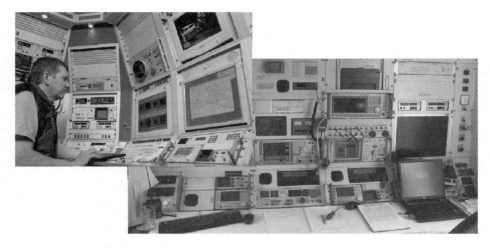

Figure IV.2   Analysis of the radio frequency spectrum following the ITU (International Telecommunication Union) guidelines in professional radio surveillance. During these procedures powerful receivers are subjected to practical daily operating situations and then pushed to their limits and beyond. (Company photograph of Federal Austrian Telecommunications Authority.)

situations in which the parameters obtained by measurement cannot adequately describe all receiving situations of practical relevance.) For this reason, an additional well-designed practical test method is important for assessing operability under everyday conditions.

Here, refer to the discussion of the second-order intercept point (Section III.9.8) of a HF radio receiver tested in the laboratory as outlined in Table IV.1. The combination of excitation signals was always chosen so that the most problematic areas are involved (these are the intermodulation excitation signals in the MW band and the SW broadcast bands, since the strongest receive signal levels are expected in these frequency regions). The specimen used (not at all favourably priced) made it necessary to provide additional information of the type shown in Table IV.1. It can be seen that regions, in which the quite impressive $IP2$ of 60 dBm drops to 30 dBm or even 20 dBm dominate. (The preselection in the unit tested is dimensioned so inadequately that there are intermediate ranges with high sum products, since the properties of a mixer are not entirely dependent on the operating frequency.) At this point, the measuring procedure may be extended almost infinitely in order to cover a large portion of the possible frequency combinations according to Equation III.27

$$f_{IM2} = f_1 + f_2$$

and to Equation III.28

$$f_{IM2} = f_2 - f_1$$

As a result one may find a point in the receiving range with an $IP2$ even below the 20 dBm already mentioned. Differences of this magnitude are not the rule but emphasize one of the problems: *It makes no sense* whatever *to assume* (on the basis of one's feelings) *any*

Table IV.1   Second order intercept points for one and the same receiver

Receive frequency (MHz)	Excitation signal at	IP2 (dBm)
7	0.5 MHz (+) 6.5 MHz	60
7	9 MHz (−) 2 MHz	44*
(7	12 MHz (−) 5 MHz	20)
14	5 MHz (+) 9 MHz	54
14	15 MHz (−) 1 MHz	30
21	7 MHz (+) 14 MHz	57*
21	9.5 MHz (+) 11.5 MHz	57

*The value decreases substantially at higher frequencies of the excitation signals. (See the values in brackets.)

good or bad properties on the basis of any single known or proven parameter. Helpful in this case are practical tests with the antenna, since these mercilessly reveal any anomalies or problematic issues possibly not even detected before, *provided that these really cause problems* (unless the propagation conditions prevent this).

IV.3   Information Gained in Practical Operation

The statement made in Section III.12 regarding the large-signal behaviour: 'all these effects shown influence the useful signal processed often simultaneously in the receiver' must be emphasized here again. Regarding Section IV.2(c) it must be assumed that for the specialist working in intelligence-related radio reconnaissance (Fig. IV.2) or the operators in an amateur radio contest it is of little relevance whether the SNR or SIR of an urgently needed useful signal is impaired by an insufficiently rejected image frequency (Section III.5.3), by cross-modulation (Section III.10) or by second-order intermodulation. For this reason this effect, which is important for practical receiving situations, is casually described as 'ghost signals and mixing products' in the 'Table of operational *PRACTICE*' describe in detail later (Table IV.3). It must be added that to distinguish or consciously separate the individual effects, as may be desirable in the metrological evaluation, is often very difficult in the hectic environment of practical operation. Identification and classification of signals and their impact requires substantial practical knowledge. However, the experienced operator knows how to improve the situation. (Table IV.2 lists typical methods for improving the reception, provided that the radio receiver used offers the functions required. How the individual interferences affect the demodulation of certain emission classes or how these sound is described together with the receiver characteristics in Part III.)

Here another important performance criterion comes into play: ergonomy. Table IV.3 specifies:

' "The existing control options allow making important adjustments" and offer the choices

"possible", "possible quickly" or "possible very quickly".'

Table IV.2   Possibilities for improving the reception in disturbed receiving situations

Interfering effect	Remedy
Receive signal at the sensitivity limit	+ Activate the RF preamplifier
Interfering signal close to the useful receive channel / insufficient near selection (Section III.6)	+ Reduce the receive bandwidth (Section III.6.1) as far as possible + Vary the IF shift (Section I.2.2)
Blocking (Section III.8)	+ Incorporate the input attenuator + Deactivate the AGC (Section III.14) and set the required gain manually (Fig. I.5)
Intermodulation (Section III.9)	+ Incorporate the input attenuator
In-band intermodulation (Section III.9.12)	+ Incorporate the input attenuator
Cross-modulation (Section III.10)	+ Incorporate the input attenuator
Distorted sound with high volume setting	+ Reduce the volume + Deactivate the AGC and set the required gain manually
Discrete interfering sounds or whistling (Section III.5)	+ Tune the notch filter (Section I.2.2) + Vary the IF shift
Crackling noise or 'pumping' with varying input signal strength	+ Change the control time constant of AGC + Deactivate the AGC and set the required gain manually

In practical operation this provides the tester with an opportunity to enter in detail, whether or not specific control functions are easily accessible (Fig. IV.3). This is because if important control options like IF shift (Section I.2.2) of a J3E receiving path, the input attenuator, or the reversal of the sideband used for demodulation of class A1A emission has to be searched in a submenu, the time required to access this function may be so long that important information may be lost in the meantime. However, if the respective controls or push-buttons are arranged on the control panel in such a way that there is no risk of accidently altering the adjustment of other parameters (even under hectic operating conditions) such a receiver would be a top candidate for the rating 'possible very quickly'. When controlling the receiver is performed through the user interface of a control computer and not manually via control elements on the front panel, the evaluation criteria can be applied accordingly.

When the text above refers to 'important control options', this means that the important functions available on any state-of-the-art receiver are accessed differently according to the manufacturer and the model. Other specific functions or unique features should be evaluated independently and in accordance with practical demands. (This is dealt with in more detail in Section IV.5.)

Table IV.3   Table of operational *PRACTICE*

RX-specific features	Assessment		
1. Ghost signals and mixing products without activated RF preamplifier are:	Audible	Slightly audible	Not audible
2. Ghost signals and mixing products with activated RF preamplifier are:	Audible	Slightly audible	Not audible
3. So-called 'crackling' and 'pumping' caused by the AGC are:	Audible	Slightly audible	Not audible
4. The subjective impression of frequency response and selectivity of the IF filters are, compared to the reference unit:	Slightly inferior	Equal	Slightly superior
5. The reception of weak signals in the presence of very strong levels are, compared to the reference unit:	Slightly inferior	Equal	Slightly superior
6. The audio reproduction of very low signals (with lower transmit station density) is, compared to the reference unit:	Slightly inferior	Equal	Slightly superior
7. The overall AF reproduction quality with the original loudspeaker is, compared to the reference unit:	Slightly inferior	Equal	Slightly superior
8. The overall AF reproduction quality with an external loudspeaker/headphone is, compared to the reference unit:	Slightly inferior	Equal	Slightly superior
9. By activating a noise reduction function the sound reproduction is, compared to the reference unit:	Slightly inferior	Equal	Slightly superior

General features	Assessment		
1. The existing control options make important adjustments:	Possible	Possible quickly	Possible very quickly
2. In a quiet environment the (ventilator) noise produced by the unit is:	Audible	Slightly audible	Not audible

Figure IV.3 It is not surprising that in the early days of radio reception one placed value on simple operation and high reproduction quality (like the benefits of high-quality headphones in this historic advertisement).

IV.3.1 Help of a Reference Unit

The most interesting comparison is the unbiased evaluation of a unit with an existing unit by simply listening to the demodulated signal of speech and music. This is true for two reasons:

(a) It becomes obvious that with average radio receivers using similar IF bandwidths there is no difference in more than 85% of all receiving situations.
(b) If any equipment is said to have a certain behaviour in a defined receiving situation it is only fair to indicate how another receiver (probably a reference unit) behaves under comparable circumstances.

Anyone interested is encouraged to perform such a comparison and will be surprised how small the differences are!

Very deliberately the words 'slightly inferior / equal / slightly superior' were chosen for the outcome of the comparison. A conventional solid receiver is used for those comparisons in which the respective line in the 'Table of operational *PRACTICE*' ends with the words 'to the reference unit'. It is important that the same reference unit is used in all test series. This prevents subjective impressions, protects against prejudices (positive and negative since everyone has preferences) as there are only minor differences between the two units operating simultaneously. Both units are connected to the *same* antenna via a coaxial switch.

IV.3.2 A Fine Distinction is Hardly Possible or Necessary

Owing to the reasons described, a finer differentiation than the three categories offered in the 'Table of operational *PRACTICE*' seems illusory. On the contrary, a distinction in finer steps would allow more leeway for subjectivity. (Remember that a more or less transparent comparison with other units should be feasible even under different propagation conditions.)

The statement 'weakly audible' is most justified when an effect influences the useful signal only to a slight extent or very seldom (probably not more than in 10% of test conditions provoked) within a testing period.

IV.4 Interpretation (and Contents of the 'Table of operational *PRACTICE*')

Table IV.3, compiled to take account of the considerations and experiences described, comprises two parts.

The features in the *upper part of the table* relate to the behaviour of the radio receiver under certain receiving antenna conditions. The 'very low signals' referred to in item 6 are signals that are just above the noise level. The reference unit, which should remain the same throughout all test series, can be regarded as virtually a standard. A radio receiver with, for example, a 3 dB lower noise figure may not produce as clear a sound as a unit having a slightly higher noise figure, as already mentioned in Section IV.2(d). From a technical point of view it must be added that if the installation position of the loudspeaker has an attenuating effect on certain audible sound frequency ranges, this narrows the bandwidth much the same as a filter effect, and thus reduces the noise perceived. (If the signal information is supplied to a decoder instead of the human ear, the filter effect of the loudspeaker has no influence on processing the signal, so that here again the measured value is a valuable parameter.)

The noise reduction functions referred to in item 9 refer primarily to the noise attenuation achieved by digital signal processing (DSP), provided that the specimen offers this possibility. A true gain in quality is achieved only by better and distortion-free sound reproduction under the same receiving conditions. Of particular interest is the behaviour of the test unit in the case of fast fading. The comparison and test results should be based on the reference unit. Taking a closer look at the advertised 'magnificent innovation' for improving the reception often reveals that a positive impression is realized only with a sufficiently high signal-to-noise ratio. But with decreasing SNR or SIR, not much of the expected benefit may be left, and the physical limitations (Section III.4.1) become noticeable.

Under 'General features', the *second part of the table* gives information about the noise produced by the specimen, since the noise not only varies greatly between different products but is also rather disturbing. An evaluation criterion regarding the (simplicity of) operation is also given in this part.

There are always three choices for the assessment.

Figure IV.4 By way of example, antenna systems of this type for the intelligence-related acquisition of information feed complex signal scenarios to the receiver input. (Company photograph of
FS Antennentechnik.)

IV.4.1 The Gain in Information

The 'Table of operational *PRACTICE*' (or equivalent information) summarizes information about the behaviour of a receiver in different operational situations within narrow
bounds. This *supports* the recorded measuring results and provides information about
the interaction of various device-specific parameters when the receiver is connected to
an antenna system that is as powerful as possible (Fig. IV.4). Under real operational
conditions this will demonstrate to the listener the effects resulting from the strong and
weak points of a receiver determined by measurement. The fact, *that basic equipment
features are treated in the same way* appears to be especially important. Nevertheless,
we are speaking of (subjective) human impressions that may differ, quite contrary to
measured values!

One should not be surprised if a cost-efficient unit with a rather modest third-order intercept point (Section III.9.8) produces equally good or even better results than a receiver of a higher price category. If looking at $IP3$s alone or separated from other parameters, their significance is limited (Section III.12.2); the listening comparison is sometimes an eye-opener, when looking for a design offering a good price–performance ratio.

IV.5 Specific Equipment Details

Depending on the actual application of a radio receiver, specific design features may be of particular significance for the user. These must be evaluated separately by individual suitability tests. This concerns the following criteria and features in particular:

(a) (Sturdiness of the) mechanical equipment construction.
(b) Size of control elements and their electromechanical properties (like ease of movement and flywheel effect), space between the controls and tactility of buttons.
(c) Operator guidance and structure of submenus, that is, the ergonomics of operation. Are all functions repeatedly used in the intended application directly accessible by means of dedicated control elements? (For example, how sensitive is frequency tuning without changing the increment repeatedly?)
(d) Readability and contrast of displays, instruments and lettering. Is the unit equipped with an adjustable foot, allowing the viewing angle to be adapted to the setup site?
(e) Computer control; possibility of integration into a LAN network and remote-controlled operation.
(f) Power consumption and/or life of the rechargeable battery, as well as weight and dimensions of portable units.
(g) Structure of the user manual and other equipment documentation.
(h) Functionality, design and suitability of optional functions or automated (measuring) routines, for example, in analysis receivers (Section II.4.4) for the intended use of the equipment at the user site.

References

[1] Gunnar Kautza: Moderne Fernmeldekommunikation auf Marineschiffen (Modern Telecommunications on Marine Craft); manuscripts of speeches at the RADCOM Hamburg 2007, pp. 1–20

[2] Frank Sichla, Ralf Rudersdorfer: So misst die 'funk' – Messvorschrift für KW-Empfänger, -Sender, -Transceiver-Empfangs- und Sendeteile (The Way 'funk' Measures – Measuring Instructions for Receiving and Transmitting Components of SW Receivers, Transmitters and Transceivers); funk 2/2002, pp. 68–71; ISSN 0342-1651

[3] Frank Sichla, Ralf Rudersdorfer: So misst die 'funk' – Messvorschrift für FM-Empfänger, -Sender, -Transceiver-Empfangs- und Sendeteile (The Way 'funk' Measures – Measuring Instructions for Receiving and Transmitting Components of FM Receivers, Transmitters and Transceivers); funk 3/2002, pp. 68–70; ISSN 0342-1651

Further Reading

Reinhard Birchel: Mechanische Filter (Mechanical Filters); funk 4/2002, pp. 34–36; ISSN 0342-1651

Rainer Bott, Jens Pöhlsen: Marinekommunikation im technologischen Wandel – Neue Konzepte für die vernetzte Operationsführung (Marine Communications under Technological Change – New Concepts of Network-Based Operations); MARINEFORUM 6/2007, pp. 12–15; ISSN 0172-8547

Robert Matousek: Fußball-Europameisterschaft 2008 – auch technisch eine Herausforderung (European Soccer Championship 2008 – a Technical Challenge); Neues von Rohde&Schwarz (Rohde&Schwarz News) IV/2008, pp 74–76; ISSN 0548-3093

Hermann Weber, Heinrich Otruba, publisher: Funküberwachung (Radio Surveillance); TELELETTER der obersten österreichischen Fernmeldebehörde und der Telekom-Control GmbH (TELELETTER issued by the supreme Telecommunications Authority of Austria and by Telekom-Control Ltd.) 3/1998, pp. 3–6

V

Concluding Information

V.1   Cascade of Noisy Two-Ports (Overall Noise Performance)

With a ladder network of several noisy two-ports the noise power is summed. The noise figure of the entire system can be derived from the noise figure (Section III.4.2) of the individual discrete two-ports in series and their power gain figure. (With passive two-ports like attenuation pads, feeder lines, etc. the attenuation figure can be viewed as a negative gain figure.)

Here, the use of the *available power* gain figure [1] is essential since especially noise-optimized stages lack general matching (Section III.3) or have no uniformly constant wave impedance. The available power gain figure is the difference between the available power level at the output of the respective two-port and that of the upstream element supplying the signal. The available power of a signal source is generally

$$P_{avl} = \frac{V_{EMF}^2}{4 \cdot R_i} \tag{V.1}$$

where

P_{avl} = available power at the output, in W
V_{EMF} = rms value of the source voltage (Section III.4.7), in V
R_i = internal resistance, in Ω

The higher the amplification of the first element of the ladder network, the smaller is the noise contribution of the second two-port (as seen from the input side):

$$F_{dB\,tot} = 10 \cdot \lg \left(10^{\frac{F_{dB\,1}}{10}} + \frac{10^{\frac{F_{dB\,2}}{10}} - 1}{10^{\frac{G_{dB\,1}}{10}}} + \frac{10^{\frac{F_{dB\,3}}{10}} - 1}{10^{\frac{G_{dB\,1} + G_{dB\,2}}{10}}} + \cdots \right.$$

$$\left. + \frac{10^{\frac{F_{dB\,n}}{10}} - 1}{10^{\frac{G_{dB\,1} + G_{dB\,2} + G_{dB\,3} + \cdots + G_{dB\,n-1}}{10}}} \right) \tag{V.2}$$

Radio Receiver Technology: Principles, Architectures and Applications, First Edition. Ralf Rudersdorfer.
© 2014 Ralf Rudersdorfer. Published 2014 by John Wiley & Sons, Ltd.

where

$$F_{dB\,tot} = \text{total noise figure, in dB}$$
$$F_{dB\,1}, F_{dB\,2}, F_{dB\,3}, \ldots, F_{dB\,n} = \text{noise figure of the respective two-port, in dB}$$
$$G_{dB\,1}, G_{dB\,2}, G_{dB\,3}, \ldots, G_{dB\,n-1} = \text{power gain figure of the respective two-port, in dB}$$

For example, let us assume that a HF receiver with good large-signal immunity and with characteristic values according to Table III.5 follows an upstream low-noise receiving converter for detecting weakest signals in the UHF range (Fig. V.1). The UHF/HF receiving converter has a noise figure of 2.7 dB with an available power gain figure of 18 dB. (The converter can be regarded as an input stage converting the signal to a lower IF by mixing (Section V.4) after the low-noise amplification. The IF corresponds to the receive frequency of the subsequent HF radio receiver performing the narrow-band selection and demodulation. If viewed as a whole, it virtually forms a heterodyne receiver (Fig. I.6)

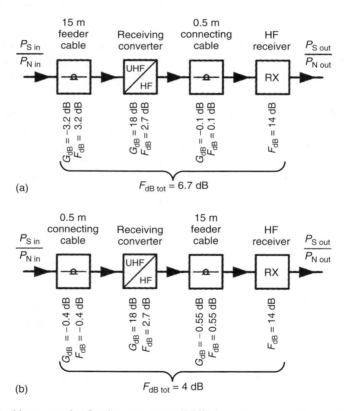

Figure V.1 Ladder network of noisy two-ports. While in arrangement (a) the attenuation figure of the long feeding line carrying the high receive frequency makes a large contribution to the total noise figure, arrangement (b) brings an increase in sensitivity. One reason for the low noise is that the low noise amplification is close to the beginning of the ladder. Another reason is that, at a lower operating frequency, the cables cause less attenuation. It is therefore advantageous to place long feeding lines behind the frequency-converting UHF/HF receiving converter. (In this example this has relatively little effect, since it is already arranged behind the amplifying element.)

with an additional intermediate frequency. The advantage of such a system is that the connecting cable between the converter and the receiver carries only the down-mixed signal of low frequencies, despite the high receive frequencies. This reduces the losses due to line attenuation. Furthermore, it provides the opportunity to expand the receive frequency range of the usually high-end receiver at relatively low costs since the components of the existing down-stream radio receiver can be used. In this example, a 15 m coaxial cable with assembled coaxial connectors is used for the connection to the receiving antenna at the receive frequency in the UHF range. The coaxial cable with plugs has an attenuation figure of 3.2 dB (Table V.1). When placing the components in series as shown in Figure V.1(a), the attenuation of the additionally required 500 mm connecting cable between the receiving converter and the HF receiver is only ~0.1 dB because of the low frequency. The total noise figure of the arrangement can be calculated with Equation (V.2)

$$F_{\text{dB tot}} = 10 \cdot \lg \left(10^{\frac{3.2\,\text{dB}}{10}} + \frac{10^{\frac{2.7\,\text{dB}}{10}} - 1}{10^{\frac{-3.2\,\text{dB}}{10}}} + \frac{10^{\frac{0.1\,\text{dB}}{10}} - 1}{10^{\frac{-3.2\,\text{dB} + 18\,\text{dB}}{10}}} + \frac{10^{\frac{14\,\text{dB}}{10}} - 1}{10^{\frac{-3.2\,\text{dB} + 18\,\text{dB} + (-0.1\,\text{dB})}{10}}} \right)$$

$$= 6.7\,\text{dB}$$

This represents a remarkable improvement of the sensitivity compared with the noise figure of 14 dB of the initial sensitivity (Section III.4) of the HF receiver. The value is

Table V.1   Characteristic values of cascading two-ports for the examples described

Property	Characteristic value
HF receiver	
Noise figure	$F_{\text{dB}} = 14\,\text{dB}$
Third order intercept point*	$IP3 = 18\,\text{dBm}$
Effective intercept point*	$IP3_{\text{eff}} = 4\,\text{dBm}$
Second order intercept point	$IP2 = 66\,\text{dBm}$
UHF/HF receiving converter	
Noise figure	$F_{\text{dB}} = 2.7\,\text{dB}$
Power gain figure	$G_{\text{dB}} = 18\,\text{dB}$
Third order intercept point*	$IP3 = 10\,\text{dBm}$
Effective intercept point*	$IP3_{\text{eff}} = 7.3\,\text{dBm}$
Second order intercept point	$IP2 = 35\,\text{dBm}$
15 m feeding line with coaxial connectors	
Attenuation figure for UHF receive frequency	$a = 3.2\,\text{dB}$
Attenuation figure for the receive frequency of the HF receiver	$a = 0.55\,\text{dB}$
500 mm feeding line with coaxial connectors	
Attenuation figure for UHF receive frequency	$a = 0.4\,\text{dB}$
Attenuation figure for the receive frequency of the HF receiver	$a \approx 0.1\,\text{dB}$

*With at least 30 kHz frequency spacing of the interfering carrier/carriers.

low even though the 15 m long feeding cable to the amplifying low-noise converter and its high cable attenuation due to the receive frequency are fully considered in the calculation.

If it is possible to place the UHF/HF receiving converter in a weather protection housing in close vicinity to the antenna feeding point, the actual advantages regarding the receiving sensitivity can be fully utilized. This reduces the attenuation figure of the coaxial cable [2] with the frequency used by the factor

$$ x \approx \sqrt{\frac{f_{up}}{f_{low}}} \tag{V.3} $$

where
   $x =$ factor by which the attenuation figure (in dB!) of the coaxial cable varies
          between two different frequencies, dimensionless
 $f_{up} =$ upper operating frequency, in Hz
$f_{low} =$ lower operating frequency, in Hz

and it also reduces the attenuation figure of the 15 m coaxial cable. The improved conditions are shown in Fig. V.1(b) in greater detail. A new calculation of the total noise figure of the noise-optimized receiving system results in the reduced noise figure of

$$ F_{dB\,tot} = 10 \cdot \lg \left(10^{\frac{0.4\,dB}{10}} + \frac{10^{\frac{2.7\,dB}{10}} - 1}{10^{\frac{-0.4\,dB}{10}}} + \frac{10^{\frac{0.55\,dB}{10}} - 1}{10^{\frac{-0.4\,dB + 18\,dB}{10}}} + \frac{10^{\frac{14\,dB}{10}} - 1}{10^{\frac{-0.4\,dB + 18\,dB + (-0.55\,dB)}{10}}} \right) $$

$$ = 4\,dB $$

Conditions similar to those described above are achieved even when using a low-noise preamplifier instead of the frequency reducing receiving converter. Only the line loss along the cables before and behind the amplifier produces the full effect, according to the common frequency position. Replacing the UHF/HF receiving converter in the receiving system shown in Fig. V.1 by a low-noise amplifier (LNA) and with all other parameters kept the same, the total noise figure for (a) is then 6.8 dB. The arrangement shown in Fig. V.1(b) reduces the value of $F_{dB\,tot}$ to 4.7 dB. However, the actual radio receiver must be capable of processing the high receive frequencies. The higher the receive frequency or the cable loss, the more effective is the concept of using a frequency converting receiving converter.

This reflects the general rule for reducing the noise in favour of a higher sensitivity: The first component in a chain should have a high amplification with little inherent noise and, if possible, the elements at the end of the chain should have a low amplification [3] or attenuating elements in order to obtain advantageous overall characteristics.

V.2   Cascade of Intermodulating Two-Ports (Overall Intermodulation Performance)

In a reaction-free ladder network of several intermodulating two-ports the intermodulation products (Fig. III.59) add up in the worst case. Relative to the input of each discrete two-port and their individual gain figures the overall intercept point can be calculated on the basis of the input intercept points (Section III.9.8). This allows estimating the intermodulation immunity (Section III.9.6) of the overall system. (With passive two-ports like attenuation pads, feeding lines, etc. the attenuation figure can be regarded as a negative gain figure. As long as the passive intermodulation is negligible, a correspondingly high IPIP should be used in the calculation.) *This assumes correct power matching* (Section III.3) *of the individual elements connected in order to obtain correct results.* With the limiting conditions regarding correctly measured intercept points for a certain frequency spacing of the interfering carriers (for example 30 kHz throughout) as described in Section III.9, the following applies to the elements in the chain.

V.2.1 Overall Third-Order Intercept Point

$$IP3_{\text{tot}} = -10 \cdot \lg \left(10^{\frac{-IPIP3_1}{10}} + 10^{\frac{G_{\text{dB}\,1} - IPIP3_2}{10}} + 10^{\frac{G_{\text{dB}\,1} + G_{\text{dB}\,2} - IPIP3_3}{10}} + \ldots \right.$$

$$\left. + 10^{\frac{G_{\text{dB}\,1} + G_{\text{dB}\,2} + \ldots + G_{\text{dB}\,n-1} - IPIP3_n}{10}} \right) \qquad (V.4)$$

where

$$IP3_{\text{tot}} = \text{total intercept point of third order, in dBm}$$
$$IPIP3_1, IPIP3_2, IPIP3_3, \ldots, IPIP3_n = \text{input intercept point of third order of the respective two-port, in dBm}$$
$$G_{\text{dB}\,1}, G_{\text{dB}\,2}, G_{\text{dB}\,3}, \ldots, G_{\text{dB}\,n-1} = \text{power gain figure of the respective two-port, in dB}$$

The following considerations refer to the same example of a noise-optimized receiving system as described in Section V.1. The sensitive UHF/HF receiving converter has an *IP*3 (Section III.9.8) of 10 dBm relative to the input and frequency spacing of the intermodulating excitation signals of at least 30 kHz. The characteristic values of the HF receiver used are known from Table III.5 and are listed again in Table V.1 in summary form for the example used. The block diagram of the receiving system is shown in Figure V.2.

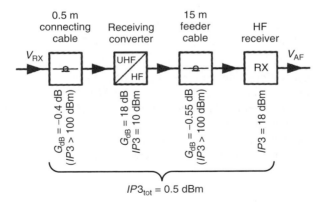

Figure V.2 Ladder network of intermodulating two-ports for the example described having the characteristic values listed in Table V.1 – specifically regarding the behaviour of third-order intermodulation.

Based on the amplification in front of the HF receiver, Equation (V.4) leads to an overall third-order intercept point of

$$IP3_{tot} = -10 \cdot \lg \left(10^{\frac{-100\text{ dBm}}{10}} + 10^{\frac{-0.4\text{ dB} - 10\text{ dBm}}{10}} + 10^{\frac{-0.4\text{ dB} + 18\text{ dB} - 100\text{ dBm}}{10}} \right.$$

$$\left. + 10^{\frac{-0.4\text{ dB} + 18\text{ dB} + (-0.55\text{ dB}) - 18\text{ dBm}}{10}} \right) = 0.5\text{ dBm}$$

This value is well suited for many ranges of UHF reception, even under difficult reception situations. This is especially the case in view of the sensitivity of the receiving system, with an overall noise figure of only 4 dB and the small frequency spacing of only 30 kHz between the interfering signals. There are almost no strong interfering signals simultaneously in the high receive frequency ranges (UHF/SHF) so close to the useful channel.

(Another method exists for calculating the overall intercept points, which considers the IPs of the various stages as voltages across a 50 Ω impedance. These voltages must be added vectorially. This leads to a somewhat lower overall intercept point compared with the calculation using the RF power levels described above. In practice, however, this method of calculation is of no significance. The method using the RF power levels has become the standard for developing cascading systems.)

V.2.2 Overall Second-Order Intercept Point

$$IP2_{tot} = -20 \cdot \lg \left(10^{\frac{-IPIP2_1}{20}} + 10^{\frac{G_{\text{dB}1} - IPIP2_2}{20}} + 10^{\frac{G_{\text{dB}1} + G_{\text{dB}2} - IPIP2_3}{20}} + \ldots \right.$$

$$\left. + 10^{\frac{G_{\text{dB}1} + G_{\text{dB}2} + \ldots + G_{\text{dB}n-1} - IPIP2_n}{20}} \right) \qquad (V.5)$$

where

$$IP2_{tot} = \text{total intercept point of second order, in dBm}$$
$$IPIP2_1, IPIP2_2, IPIP2_3, \ldots, IPIP2_n = \text{input intercept point of second order of the respective two-port, in dBm}$$
$$G_{\text{dB}1}, G_{\text{dB}2}, G_{\text{dB}3}, \ldots, G_{\text{dB}n-1} = \text{power gain figure of the respective two-port, in dB}$$

The following considerations refer to the example of a noise-optimized receiving system described in Sections V.1 and V.2.1. The sensitive UHF/HF receiving converter has an *IP2* (Section III.9.8) of 35 dBm relative to the input. The characteristic values of the HF receiver used are known from Table III.5 and are listed again in Table V.1 in summary form for the example used. The block diagram of the receiving system is shown in Figure V.3. On this basis, Equation (V.5) leads to an overall second order intercept point of

$$IP2_{tot} = -20 \cdot \lg \left(10^{\frac{-200\text{ dBm}}{20}} + 10^{\frac{-0.4\text{ dB} - 35\text{ dBm}}{20}} + 10^{\frac{-0.4\text{ dB} + 18\text{ dB} - 200\text{ dBm}}{20}} \right.$$

$$\left. + 10^{\frac{-0.4\text{ dB} + 18\text{ dB} + (-0.55\text{ dB}) - 66\text{ dBm}}{20}} \right) = 34\text{ dBm}$$

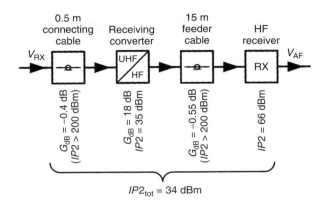

Figure V.3 Ladder network of intermodulating two-ports for the example described having the characteristic values listed in Table V.1 – specifically regarding the behaviour of second-order intermodulation.

Furthermore, the receiving system is relatively immune against IM2 in relation to the high receive frequency range (UHF/SHF).

In practical designs the discrete two-ports arranged in series often have no wide-band matching, for example, outside the passband of selective elements. Especially for IM2 this causes real differences in the results compared with Equation (V.5), owing to the wide frequency spacing of the signals (Section III.9.2). For both the interfering carrier frequencies and the frequency combinations causing the IM products the method of calculation assumes a real termination of the individual elements of the ladder network equal to the internal resistance (therefore impedance matching).

With regard to intermodulation (non-linear distortions) it becomes clear that in order to cope with these interfering effects it is advantageous to have a low amplification at the beginning of the ladder and the elements at the end have a higher amplification [3]. When considering the noise interferences (Section V.1) it is obvious that just the opposite system design would reduce the noise: The high (low-noise) amplification at the beginning produces a favorable overall noise behaviour. *This shows that two-port parameters (that is, their individual characteristic values) must always be adapted to the requirements of the actual receiving situation.* (For optimizing the reception in different signal situations it is therefore ideal to avoid the influence of the various cascade elements, for example using switchable bypasses.)

V.2.3   Computer-Aided Calculations

Modern software is available for PC-based methods of examining cascaded systems. Several versions are available free of charge. Especially to be emphasized is the program AppCAD [4] (Fig. V.4) also used in developing laboratories in industry. Among other features, AppCAD allows the complete evaluation of receiving systems and the quick and easy determination of the overall noise figures and intercept points for active and passive

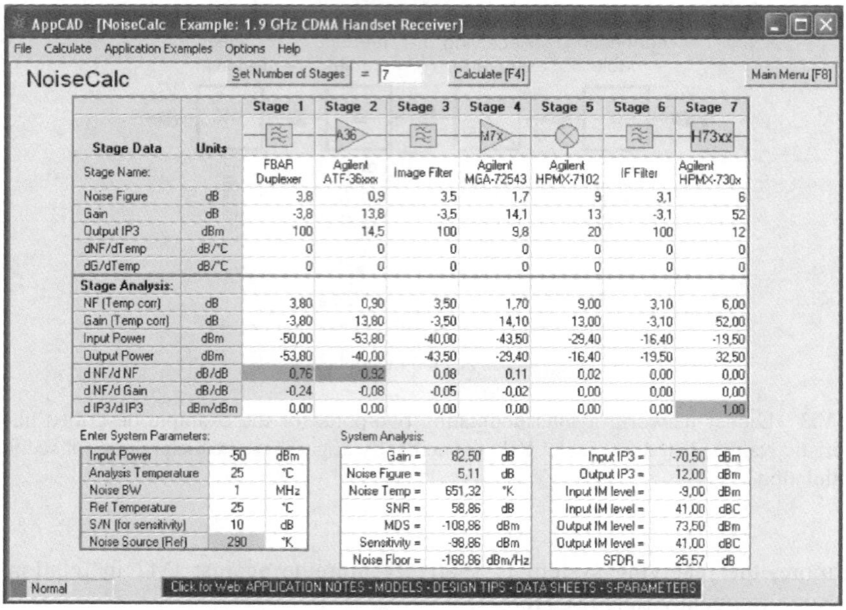

AppCAD [NoiseCalc Example: 1.9 GHz CDMA Handset Receiver]

File Calculate Application Examples Options Help

NoiseCalc   Set Number of Stages = 7   Calculate [F4]   Main Menu [F8]

Stage Data	Units	Stage 1	Stage 2	Stage 3	Stage 4	Stage 5	Stage 6	Stage 7
Stage Name:		FBAR Duplexer	Agilent ATF-36xxx	Image Filter	Agilent MGA-72543	Agilent HPMX-7102	IF Filter	Agilent HPMX-730x
Noise Figure	dB	3.8	0.9	3.5	1.7	9	3.1	6
Gain	dB	-3.8	13.8	-3.5	14.1	13	-3.1	52
Output IP3	dBm	100	14.5	100	9.8	20	100	12
dNF/dTemp	dB/°C	0	0	0	0	0	0	0
dG/dTemp	dB/°C	0	0	0	0	0	0	0
Stage Analysis:								
NF (Temp corr)	dB	3.80	0.90	3.50	1.70	9.00	3.10	6.00
Gain (Temp corr)	dB	-3.80	13.80	-3.50	14.10	13.00	-3.10	52.00
Input Power	dBm	-50.00	-53.80	-40.00	-43.50	-29.40	-16.40	-19.50
Output Power	dBm	-53.80	-40.00	-43.50	-29.40	-16.40	-19.50	32.50
d NF/d NF	dB/dB	0.76	0.92	0.08	0.11	0.02	0.00	0.00
d NF/d Gain	dB/dB	-0.24	-0.08	-0.05	-0.02	0.00	0.00	0.00
d IP3/d IP3	dBm/dBm	0.00	0.00	0.00	0.00	0.00	0.00	1.00

Enter System Parameters:

Input Power	-50	dBm
Analysis Temperature	25	°C
Noise BW	1	MHz
Ref Temperature	25	°C
S/N (for sensitivity)	10	dB
Noise Source (Ref)	290	°K

System Analysis:

Gain =	82.50	dB
Noise Figure =	5.11	dB
Noise Temp =	651.32	°K
SNR =	58.86	dB
MDS =	-108.86	dBm
Sensitivity =	-98.86	dBm
Noise Floor =	-168.86	dBm/Hz

Input IP3 =	-70.50	dBm
Output IP3 =	12.00	dBm
Input IM level =	-9.00	dBm
Input IM level =	41.00	dBC
Output IM level =	73.50	dBm
Output IM level =	41.00	dBC
SFDR =	25.57	dB

Normal   Click for Web: APPLICATION NOTES - MODELS - DESIGN TIPS - DATA SHEETS - S-PARAMETERS

Figure V.4 Input screen of the AppCAD program [4] for the simplified calculation and analysis of system parameters.

circuit elements arranged in series. (This again assumes appropriate matching between the discrete elements in order to obtain correct results.)

V.3 Mathematical Description of the Intermodulation Formation

Each continuous transfer characteristic of an active non-memory two-port can be described by a general polynomial in the form of a series expansion [5]:

$$v_{out} = v_{out\,0} + a \cdot v_{in} + b \cdot v_{in}^2 + c \cdot v_{in}^3 + d \cdot v_{in}^4 + e \cdot v_{in}^5 + \ldots \qquad (V.6)$$

where

v_{out} = output voltage, in V

v_{in} = input voltage, in V

With an input voltage of $0\,V$ the output voltage is $v_{out\,0}$. The term $v_{out} = a \cdot v_{in}$ contains the ideally linear amplification a and represents the linear portion of the characteristic curve. All other terms with b, c, d, e, ... enable the generally applicable but concrete description of non-linear amplification of the limiting/distorting curvature in the transfer characteristic (Fig. III.56) together with the respective evaluation factors and algebraic signs.

Feeding to a two-port the general transfer function as shown above with the sum of two sinusoidal signals differing in both amplitude (v_1 and v_2) and frequency ω_1 and ω_2,

$$v_{in} = v_1 \cdot \cos \omega_1 t + v_2 \cdot \cos \omega_2 t \qquad (V.7)$$

where

v_1 = amplitude of signal 1, in V
v_2 = amplitude of signal 2, in V
ω_1 = angular frequency of signal 1, in rad/s
ω_2 = angular frequency of signal 2, in rad/s
t = time considered, in s

this signal undergoes linear amplification to

$$v_{out} = a \cdot (v_1 \cdot \cos \omega_1 t + v_2 \cdot \cos \omega_2 t) \qquad (V.8)$$

These two signals can be regarded either as one useful and one interfering signal or as two interfering signals (interfering carriers).

V.3.1 Second-Order Intermodulation

The term $b \cdot v_{in}^2 = b \cdot (v_1 \cdot \cos \omega_1 t + v_2 \cdot \cos \omega_2 t)^2$ is the so-called *quadratic term*. It is responsible for a direct voltage component and for the generation of the second harmonic with the frequencies $2\omega_1$ and $2\omega_2$:

$$\tfrac{1}{2} \cdot b \cdot v_1^2 \cdot \cos 2\omega_1 t \quad \text{and} \quad \tfrac{1}{2} \cdot b \cdot v_2^2 \cdot \cos 2\omega_2 t \qquad (V.9)$$

and the 'notorious' *intermodulation products of second order* (IM2)

$$b \cdot v_1 \cdot v_2 \cdot \cos(\omega_1 - \omega_2)t \quad \text{and} \quad b \cdot v_1 \cdot v_2 \cdot \cos(\omega_1 + \omega_2)t \qquad (V.10)$$

One can see immediately from Equation (V.10): These signal products occur at the sum and difference frequencies of the two interfering carriers (if both signals in Equation (V.7) are viewed as interfering signals). Initially there were no signals at these frequencies. This suggests that they are produced by the non-linearity of the receiver transfer characteristic. Furthermore, one can see that these interfering products are linearly proportional to both the signal with amplitude v_1 and the signal with amplitude v_2. When increasing one interference carrier by x dB, the interference product rises by the same amount. When increasing both interference carriers, the interference products increase by the factor $v_1 \cdot v_2$, that is, by the sum of the increases in dB. If the two interference carriers are equal ($v_1 = v_2 = v$), which is usually the case in measurement engineering, then the interference products follow v^2 (quadratic term!), or change by double the dB amount (the dB figure).

(The square term of a characteristic curve forces the developer of radio receivers to use input bandpass filters with a suboctave bandwidth. Only filters having an upper cutoff frequency of less than twice the lower cutoff frequency can prevent the formation of the second harmonic and of second order IM products in the preamplifier and the mixer. In the VHF/UHF range these conditions are usually met by tailoring the necessary frontend selection to the useful frequency band.)

V.3.2   Third-Order Intermodulation

The term $c \cdot v_{in}^2 = c \cdot (v_1 \cdot \cos \omega_1 t + v_2 \cdot \cos \omega_2 t)^3$ is called the *cubic term*. After trigonometric conversion, for the frequency ω_1 this results in

$$\tfrac{3}{4} \cdot c \cdot v_1^3 \cdot \cos \omega_1 t + \tfrac{3}{2} \cdot c \cdot v_1 \cdot v_2^2 \cdot \cos \omega_1 t \qquad (V.11)$$

and for the frequency ω_2

$$\tfrac{3}{4} \cdot c \cdot v_2^3 \cdot \cos \omega_2 t + \tfrac{3}{2} \cdot c \cdot v_1^2 \cdot v_2 \cdot \cos \omega_2 t \qquad (V.12)$$

The signal components are of particular interest. Examining the original signal with frequency ω_1 as the useful signal, one can see from Equation (V.8) that two other components of the same frequency are superimposed on the proportionally amplified signal. If c has a negative sign which is normally the case, the term according to Equation (V.11) causes a *signal reduction. This is the effect known as limitation.* For a linearly increasing level of the input signal, the output signal will increase less steeply from a certain point onward. This means that there is a causal relation between odd-numbered order intermodulation and limitation. This is true in both directions: The limitation of a signal mixture inevitably causes intermodulation of odd order (Fig. III.61).

However, even more signal components are hidden in the cubic term:

$$\tfrac{3}{4} \cdot c \cdot v_1^2 \cdot v_2 \cdot \cos(2\omega_1 - \omega_2)t \qquad (V.13a)$$

$$\tfrac{3}{4} \cdot c \cdot v_1 \cdot v_2^2 \cdot \cos(2\omega_2 - \omega_1)t \qquad (V.13b)$$

The signal products from Equations (V.13a) and (V.13b) occur with a separation $|\omega_1 - \omega_2|$ above and below the frequencies ω_1 and ω_2. These are the dangerous third-order intermodulation products (IM3) because they are very close to the critical interference carriers so that a sufficient suppression is hardly possible by the usual selection methods. (Very narrow-band tunable multi-circuit preselectors (Section III.11) can help in this situation.)

Under the condition that $v_1 = v_2 = v$, which is common in measurement engineering, IM3 products are influenced by the third power of the interference voltage. An increase in the interference carrier level by 10 dB causes an increase in the IM3 products by 30 dB. If the interference carriers are not equal, for example due to a preselector (Section III.11.1), the amplitude of an IM3 product follows the further separated interference carrier linearly but follows the closer interference carrier quadratically. This suggests, *that preselection is suitable for sufficiently reducing IM3 products only if it can lower the amplitude of both interference carriers involved.*

The IM3 interference products

$$\tfrac{3}{4} \cdot c \cdot v_1^2 \cdot v_2 \cdot \cos(2\omega_1 + \omega_2)t$$

$$\tfrac{3}{4} \cdot c \cdot v_1 \cdot v_2^2 \cdot \cos(2\omega_2 + \omega_1)t \qquad (V.13c)$$

at the sum frequencies are usually of minor importance for the interference immunity as are the third order harmonics also arising.

$$\frac{1}{4} \cdot c \cdot v_1^3 \cdot \cos 3\omega_1 t + \frac{1}{4} \cdot c \cdot v_2^3 \cdot \cos 3\omega_2 t \tag{V.13d}$$

Caution: Here, the terms $(2\omega_1 - \omega_2)$ and $(2\omega_2 - \omega_1)$ represent two discrete frequencies of the respective signal components. This does NOT imply that intermodulation is caused by the second harmonic minus the fundamental wave due to mixing! This erroneous interpretation often leads to the false conclusion that the harmonics of interference signals or of the measuring carriers are responsible for the generation of IM3. This is not correct. Proof is that the second harmonics are caused exclusively by even-numbered polynomial terms, but not by the odd orders responsible for intermodulation.

V.3.3   Other Terms in the Transfer Characteristic Polynomial

The mathematical processes described above for the generation of intermodulation products of the second and third order are also described in the literature [6], and [7]. These can help to gain a deeper understanding of the underlying interrelations. However, some questions remain unanswered, such as:

(a) Why do the IM3 products generated in some RF frontends not follow the cubic law?
(b) Why can IM products increase beyond the theoretically permissible limit?

The calculation of the output voltage using the transfer characteristic polynomial provides the answers to these questions. The term $d \cdot v_{in}^4 = d \cdot (v_1 \cdot \cos \omega_1 t + v_2 \cdot \cos \omega_2 t)^4$ produces products at the following frequencies in addition to the direct voltage components:

$2\omega_1, 2\omega_2$:

$$\frac{1}{2} \cdot d \cdot v_1^4 \cdot \cos 2\omega_1 t + \frac{3}{2} \cdot d \cdot v_1^2 \cdot v_2^2 \cdot \cos 2\omega_1 t +$$
$$\frac{1}{2} \cdot d \cdot v_2^4 \cdot \cos 2\omega_2 t + \frac{3}{2} \cdot d \cdot v_1^2 \cdot v_2^2 \cdot \cos 2\omega_2 t \tag{V.14}$$

$\omega_1 - \omega_2, \omega_2 - \omega_1$:

$$\frac{3}{2} \cdot d \cdot v_1^3 \cdot v_2 \cdot \cos(\omega_1 - \omega_2)t$$
$$\frac{3}{2} \cdot d \cdot v_1 \cdot v_2^3 \cdot \cos(\omega_2 - \omega_1)t \tag{V.15a}$$

$\omega_1 + \omega_2$:

$$\frac{3}{2} \cdot d \cdot v_1^3 \cdot v_2 \cdot \cos(\omega_1 + \omega_2)t$$
$$\frac{3}{2} \cdot d \cdot v_1 \cdot v_2^3 \cdot \cos(\omega_2 + \omega_1)t \tag{V.15b}$$

and finally:
$2\omega_1 - \omega_2, 2\omega_1 + \omega_2$
$3\omega_1 - \omega_2, 3\omega_1 + \omega_2, 3\omega_2 - \omega_1, 3\omega_2 + \omega_1$
$4\omega_1, 4\omega_2$

Although this part of the transfer characteristic polynomial produces weaker signal components, many more signal components with a multitude of frequency combinations are obtained. Of importance here are those with frequencies of the second harmonics (Equation V.14). These are the reason that the formation of these harmonics does not follow the quadratic law according to Equation (V.9). Such a deviation cannot be detected in Equation (V.9) by looking only at the first terms of the polynomial. This is also true for IM2 products. The components according to Equations (V.15a) and (V.15b) add to form the quadratic components Equation (V.10) and, if the modulation is high, cause a deviation from the purely quadratic law of formation.

Other discoveries also result from the term $e \cdot v_{in}^5 = e \cdot (v_1 \cdot \cos \omega_1 t + v_2 \cdot \cos \omega_2 t)^5$ which provides signal components at the frequencies:

$\omega_1, \omega_2,$
$3\omega_1, 3\omega_2,$
$5\omega_1, 5\omega_2,$
$2\omega_1 - \omega_2, 2\omega_1 + \omega_2, 2\omega_2 - \omega_1, 2\omega_2 + \omega_1,$
$3\omega_1 - 2\omega_2, 3\omega_1 + 2\omega_2, 3\omega_2 - 2\omega_1, 3\omega_2 + 2\omega_1,$
$4\omega_1 - \omega_2, 4\omega_1 + \omega_2, 4\omega_2 - \omega_1$ and $4\omega_2 + \omega_1$

The components at the frequencies of the interference carriers ω_1 and ω_2 also influence the limitation of the maximum level. The components at $3\omega_1$ and $3\omega_2$ are added (with the respective algebraic sign) to the third harmonics of the cubic component. The fifth harmonics at $5\omega_1$, $5\omega_2$ are new. Of particular importance, however, are those components that are added to the cubic components at the frequencies of the IM3 products (see Equations V.13a and V.13b):

$$5/4 \cdot e \cdot v_1^4 \cdot v_2 \cdot \cos(2\omega_1 - \omega_2)t + 15/8 \cdot e \cdot v_1^2 \cdot v_2^3 \cdot \cos(2\omega_1 - \omega_2)t$$
$$5/4 \cdot e \cdot v_1^4 \cdot v_2 \cdot \cos(2\omega_1 + \omega_2)t + 15/8 \cdot e \cdot v_1^2 \cdot v_2^3 \cdot \cos(2\omega_1 + \omega_2)t$$
$$5/4 \cdot e \cdot v_1 \cdot v_2^4 \cdot \cos(2\omega_2 - \omega_1)t + 15/8 \cdot e \cdot v_1^3 \cdot v_2^2 \cdot \cos(2\omega_2 - \omega_1)t$$

and

$$5/4 \cdot e \cdot v_1 \cdot v_2^4 \cdot \cos(2\omega_2 + \omega_1)t + 15/8 \cdot e \cdot v_1^3 \cdot v_2^2 \cdot \cos(2\omega_2 + \omega_1)t \qquad \text{(V.16)}$$

Finally there are the additional IM5 products $3\omega_1 - 2\omega_2$, $3\omega_1 + 2\omega_1$, $3\omega_2 - 2\omega_1$, $3\omega_2 + 2\omega_1$ and the signals grouped around the third harmonics.

The terms of fifth order also occur for very high modulation of a two-port. Their components at the frequencies of the IM3 products cause them to deviate from the strict cubic law of formation. In other words: Depending on the modulation depth of the two-port under examination and according to its characteristics (for example those of a diode ring mixer, a MOSFET switching mixer or a bipolar active mixer) it is possible, from a certain limit of the dynamic range onward, the IM3 products increase by only 2.5 dB or even 3.5 dB with an increase in the interference carrier levels of 1 dB each. This deviation is owing to the fifth order components. *The components of seventh or even higher odd-numbered powers can also play a theoretical and practical role* [5].

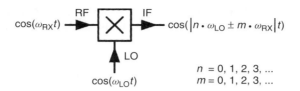

Figure V.5   Several frequency combinations of receive signal and LO injection signal are always possible for the same IF. With specific mixers of balanced or double-balanced design the harmonics mixing products are in fact weakened, but mostly still present. Good frequency planning in combination with an adapted preselection effectively counteract spurious signal reception.

V.4   Mixing and Derivation of Spurious Reception

A mixer is supplied with the receive signal to one port and the LO injection signal to the second port for heterodyning. During the mixing process, the heterodyning, the information content modulated onto the receive signal is essentially preserved, but its centre frequency with the receive frequency spectrum is shifted. This requires a circuit component with a non-linear or time-dependent voltage/current characteristic. In practice, the units are divided [8] into:

(a) Additive mixers, in which the receive signal and the LO injection signal are combined and supplied to one connector pin of the circuit component, with non-linear voltage/current characteristic (for example to the anode of a diode or to the gate of a junction field-effect transistor (JFET)).

(b) Multiplicative mixers, in which the receive signal and the LO injection signal are supplied to different connector pins of the circuit component or switching element, with time-dependent voltage/current characteristic (for example to the two gates of a dual-gate FET or a diode ring or MOSFET ring used as a switch for polar reversal).

The output signals are available at the third port and are (mathematically always) *the multiplicative result* of the two signals supplied. The output signals containing the intermediate frequency signal and the image frequencies (Section III.5.3) also carry other signal components in addition (Fig. V.5).

V.4.1   Mixing = Multiplication

The mixing process is expressed by a general polynomial from a series expansion based on the specifications made at the beginning of Section V.3:

$$i_{IF} = i_{IF\,0} + a \cdot v_{in} + b \cdot v_{in}^2 + c \cdot v_{in}^3 + \ldots \tag{V.17}$$

where
i_{IF} = output current at the IF port, in A
v_{in} = input voltage, in V

For an input voltage of 0 V the output current at the IF gate is $i_{IF\,0}$. With the sum of the two sinusoidal signals differing in both the amplitudes v_{RX} and v_{LO} as well as in the frequencies ω_{RX} and ω_{LO},

$$v_{LO} \cdot \cos \omega_{LO} t + v_{RX} \cdot \cos \omega_{RX} t \tag{V.18}$$

where
v_{LO} = amplitude of the LO injection signal, in V
v_{RX} = amplitude of the receive signal, in V
ω_{LO} = angular frequency of the LO injection signal, in rad/s
ω_{RX} = angular frequency of the receive signal, in rad/s
t = considered time, in s

a signal current is produced at the IF output of the mixer in the form described below. This contains the direct components

$$i_{IF} = i_{IF\,0} + {}^{1}\!/_{2} \cdot b \cdot v_{RX}^{2} + {}^{1}\!/_{2} \cdot b \cdot v_{LO}^{2} \tag{V.19}$$

together with the transferred or amplified useful signals supplied at their original frequency

$$a \cdot (v_{RX} \cdot \cos \omega_{RX} t + v_{LO} \cdot \cos \omega_{LO} t) \tag{V.20a}$$

$$
{}^{3}\!/_{4} \cdot c \cdot v_{RX}^{3} \cdot \cos \omega_{RX} t + {}^{3}\!/_{2} \cdot c \cdot v_{RX} \cdot v_{LO}^{2} \cdot \cos \omega_{RX} t
$$
$$
{}^{3}\!/_{4} \cdot c \cdot v_{LO}^{3} \cdot \cos \omega_{LO} t + {}^{3}\!/_{2} \cdot c \cdot v_{RX}^{2} \cdot v_{LO} \cdot \cos \omega_{LO} t \tag{V.20b}
$$

and the second harmonics of the receive signal and the LO injection signal

$$
{}^{1}\!/_{2} \cdot b \cdot v_{RX}^{2} \cdot \cos 2\omega_{RX} t \quad \text{and} \quad {}^{1}\!/_{2} \cdot b \cdot v_{LO}^{2} \cdot \cos 2\omega_{LO} t \tag{V.21}
$$

As well as the third harmonics of the receive signal and the LO injection signal

$$
{}^{1}\!/_{4} \cdot c \cdot v_{RX}^{3} \cdot \cos 3\omega_{RX} t \quad \text{and} \quad {}^{1}\!/_{4} \cdot c \cdot v_{LO}^{3} \cdot \cos 3\omega_{LO} t \tag{V.22}
$$

Furthermore, the difference and sum frequencies of second order

$$
b \cdot v_{RX} \cdot v_{LO} \cdot \cos(\omega_{LO} - \omega_{RX}) t \quad \text{and} \quad b \cdot v_{RX} \cdot v_{LO} \cdot \cos(\omega_{LO} + \omega_{RX}) t \tag{V.23}
$$

and the difference and sum frequencies of third order

$$
{}^{3}\!/_{4} \cdot c \cdot v_{RX}^{2} \cdot v_{LO} \cdot \cos(2\omega_{RX} - \omega_{LO}) t
$$
$$
{}^{3}\!/_{4} \cdot c \cdot v_{RX} \cdot v_{LO}^{2} \cdot \cos(2\omega_{LO} - \omega_{RX}) t \tag{V.24a}
$$

$$
{}^{3}\!/_{4} \cdot c \cdot v_{RX}^{2} \cdot v_{LO} \cdot \cos(2\omega_{RX} + \omega_{LO}) t
$$
$$
{}^{3}\!/_{4} \cdot c \cdot v_{RX} \cdot v_{LO}^{2} \cdot \cos(2\omega_{LO} + \omega_{RX}) t \tag{V.24b}
$$

are also found. This process continues accordingly. It can be seen that the multitude of frequencies, the harmonic mixer products, newly arising during the mixing process, follow the mathematical derivation

$$f_{IF} = |n \cdot f_{LO} \pm m \cdot f_{RX}| \qquad (V.25)$$

where
f_{IF} = (first) intermediate frequency, in Hz
n = 0, 1, 2, 3, ...
f_{LO} = LO injection frequency, in Hz
m = 0, 1, 2, 3, ...
f_{RX} = receive frequency, in Hz

V.4.2 Ambiguous Mixing Process

The use of mixer stages in receivers causes the problem of ambiguous mixing. This means that there are additional receive frequencies besides the desired setpoint receive frequency. These are clearly evident after transposing Equation (V.25) [9] to

$$f'_{RX} = \left| \frac{n}{m} \cdot f_{LO} \pm \frac{1}{m} \cdot f_{IF} \right| \qquad (V.26)$$

where
f'_{RX} = spurious receive frequency caused by harmonic mixer products, in Hz
n = 0, 1, 2, 3, ...
m = 1, 2, 3, ...
f_{LO} = LO injection frequency, in Hz
f_{IF} = (first) intermediate frequency, in Hz

Compared to the heterodyning process with $m = 1$ and $n = 1$ (fundamental wave mixing) and the image frequency reception, the additionally generated spurious receive frequencies are usually subjected to a higher conversion loss. These spurious receive frequencies are therefore noticeable in modulation with high receive levels.

In order to counteract ambiguities and thus the spurious signal reception, it must be ensured that the desired useful signal (usually given by $m = 1$, $n = 1$) is separated from all other signals arising characterized by $m \neq 1$, $n \neq 1$ by means of effective preselection before reaching the mixer stage. When using mixer stages of high linearity it can be assumed that only fundamental wave mixing with $m = 1$ is performed for the receive signal, even with high receive signal levels. If in addition the frequency plan allows the advantageous choice of $f_{LO} > f_{IF}$, the possible ambiguities are reduced to the two spurious receive frequencies

$$f'_{RX} = n \cdot f_{LO} - f_{IF} \qquad (V.27)$$

$$f'_{RX} = n \cdot f_{LO} + f_{IF} \qquad (V.28)$$

where
f'_{RX} = spurious receive frequency caused by harmonic mixer product, in Hz
  n = 1, 2, 3, ...
f_{LO} = LO injection frequency, in Hz
  f_{IF} = (first) intermediate frequency, in Hz

per harmonic of the LO injection signal.

V.5   Characteristics of Emission Classes According to the ITU RR

The emission class of a transmission (Table II.5) is identified by five characters according to [10]. The first three characters define the main characteristics and are composed of a sequence of letter, numeral, letter as listed in Table V.2. Another two letters are available for property details and the type of channel utilization in the transmission method used. (The two last digits however are rarely used.)

The bandwidth of an emission is given by three numerals and one letter and should be used as a prefix to the class of emission. The letters used are H (Hz), K (kHz), M (MHz) and G (GHz) and are inserted instead of the decimal point.

Table V.3 lists examples of classic emission classes often used together with the correct designation of their properties.

V.6   Geographic Division of the Earth by Region According to ITU RR

For the international (supra-regional) allocation of frequencies the world has been divided into three regions as shown on the map in Figure V.6. These can be described as follows in accordance with [11]:

- *Region 1* includes Europe, Africa and Arabia, excluding any of the territory of the Islamic Republic of Iran which lies outside of the limits shown in Figure V.6. It also includes the whole of the territory of Armenia, Azerbaijan, the Russian Federation, Georgia, Kazakhstan, Mongolia, Uzbekistan, Kyrgyzstan, Tajikistan, Turkmenistan, Turkey and Ukraine and the area to the north of Russian Federation.
- *Region 2* includes Greenland, North, Central and South America.
- *Region 3* includes Australia and (primarily South, Southwest and East) Asia, except the territory of Armenia, Azerbaijan, the Russian Federation, Georgia, Kazakhstan, Mongolia, Uzbekistan, Kyrgyzstan, Tajikistan, Turkmenistan, Turkey and Ukraine and the area to the north of the Russian Federation. It also includes that part of the territory of the Islamic Republic of Iran which lies inside of the boundaries shown Figure V.6.

V.7   Conversion of dB ... Levels

A simple and efficient conversion between different level specifications can be performed with the conversion tables described below. The computation effort required is limited to

Table V.2 Composition of emission class designations according to the ITU RR [10]

Modulation type of the main carrier		Signal type of the modulated main carrier		Type of information		Property details		Type of multiplex (type of multiple channel access)	
N	Non-modulated carrier	0	No modulating signal	N	No information	A	Two-state code with different number and/or duration of elements	N	No multiple use
A	Main carrier with double sideband AM	1	A single channel containing quantized or digital information	A	Telegraphy – for audio reception	B	Two-state code with equal number and/or duration of elements, without error correction	G	Code multiplexing (CDMA)
H	Main carrier with single sideband AM	2	A single channel containing quantized or digital information (utilizing a modulating subcarrier)	B	Telegraphy – for automated reception	C	Two-state code with equal number and/or duration of elements, with error correction	F	Frequency multiplexing (FDMA)
R	Main carrier with single sideband AM and reduced carrier	3	A single channel containing analog information	C	Fax	D	Four-state code; each state represents a signal element (of one or several bits)	T	Time multiplexing (TDMA)
J	Main carrier with single sideband AM and (fully) suppressed carrier	7	Two or more channels containing quantized or digital information	D	Data transmission, remote control, remote measuring	E	Four-state code; each state represents a signal element (of one or several bits)	W	Combination of frequency multiplexing (FDMA) and time multiplexing (TDMA)
B	Main carrier with independent sideband modulation	8	Two or more channels containing analog information	E	Telephony (including audio broadcasting)	F	Four-state code; each state or combination of states represents a character	X	Other cases of multiplexing
C	Main carrier with vestigial sideband modulation	9	Combination of the above cases	F	Television (video)	G	Sound in broadcasting quality (monophone)		
F	Main carrier with FM	X	Other cases	G	Combination of the above cases	H	Sound in broadcasting quality (stereophone or quadrophone)		

(continued overleaf)

Table V.2 (*continued*)

Modulation type of the main carrier	Signal type of the modulated main carrier	Type of information	Property details	Type of multiplex (type of multiple channel access)
G Phase modulation		X Other cases	J Sound in commercial quality (except the cases under K and L)	
P Emission of unmodulated pulses		X Other cases	J Sound in commercial quality (except the cases under K and L)	
K Emission of amplitude-modulated pulses			K Sound in commercial quality with audio frequency band in inverted position or with AF band divided in sections	
L Emission of width-modulated pulses			L Sound in commercial quality with separate FM signals to control the demodulated signal levels	
M Emission of phase-modulated pulses			M Black and white image	
Q Emission of pulses with the carrier angle-modulated during the pulse duration			N Colour image	
V Emission of pulses that are a combination of the above cases or of other conditions			W Combination of the above cases	
W Main carrier with a combination of the above methods			X Other cases	
X Other cases				

Table V.3 Examples of designations for emission classes often used

Class of emission	Characteristics
	A1A
A	Main carrier with double sideband AM
1	A single channel containing quantized or digital information
A	Telegraphy for audible reception
Used in Morse telegraphy with keyed carrier	
	A3E
A	Main carrier with double sideband AM
3	A single channel containing analog information
E	Telephony (including audio broadcasting)
Typically used in AM audio broadcasting	
	F2D
F	Main carrier with FM
2	A single channel containing quantized or digital information (utilizing a modulated subcarrier)
D	Data transmission, remote control, remote measuring
Typically used in digital (A)FSK modulated transmission methods	
	F3E
F	Main carrier with FM
3	A single channel containing analog information
E	Telephony (including audio broadcasting)
Typically used in FM voice radio	
	G3E
G	Phase modulation
3	A single channel containing analog information
E	Telephony (including audio broadcasting)
Typically used in VHF FM audio broadcasting with preemphasis	
	J3E
J	Main carrier with single sideband AM and (fully) suppressed carrier
3	A single channel containing analog information
E	Telephony (including audio broadcasting)
Typically used in SSB voice radio	

simple addition and subtraction. More frequently performed conversions like from dBm to dBμV and vice versa are simplified by memorizing the respective numerical value ('107' in the example dBm versus dBμV).

Tables V.4 and V.5 can be combined for converting parameter values if the K factor or the antenna gain of the receiving antenna used are known. This procedure is outlined in Section III.18.

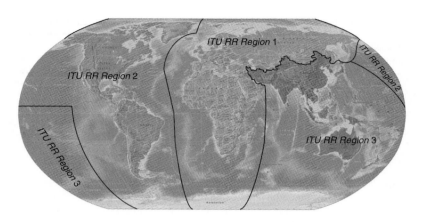

Figure V.6 Division of the world into three geographic regions according to the ITU RR [11].

V.7.1 Voltage, Current and Power Levels

Table V.4 enables simple conversion between voltage levels, current levels and power levels in a 50 Ω system.

For example, at a 50 Ω resistor a voltage level of −41 dBmV induces a power level of

$$-41\,\text{dBmV} - 77 = -118\,\text{dBW}$$

or, expressed in dBm,

$$-41\,\text{dBmV} - 47 = -88\,\text{dBm} \cong -118\,\text{dBW} + 30 = -88\,\text{dBm}$$

caused by a current level of

$$-41\,\text{dBmV} - 34 = -75\,\text{dBmA} \cong -88\,\text{dBm} + 13 = -75\,\text{dBmA}$$

flowing through a 50 Ω resistor.

Table V.4 Conversion between voltage / current / power levels in 50 Ω systems

from ↓ to →	dBW	dBm	dBV	dBmV	dBμV	dBA	dBmA	dBμA
dBW		+30	+17	+77	+137	−17	+43	+103
dBm	−30		−13	+47	+107	−47	+13	+73
dBV	−17	+13		+60	+120	−34	+26	+86
dBmV	−77	−47	−60		+60	−94	−34	+26
dBμV	−137	−107	−120	−60		−154	−94	−34
dBA	+17	+47	+34	+94	+154		+60	+120
dBmA	−43	−13	−26	+34	+94	−60		+60
dBμA	−103	−73	−86	−26	+34	−120	−60	

Table V.5 Conversion of field strength / (power) flux density levels for free-space propagation in the far field

from ↓ to →	dB(W/m²)	dB(mW/m²)	dB(mW/cm²)	dB(V/m)	dB(mV/m)	dB(µV/m)	dB(A/m)	dB(mA/m)	dB(µA/m)	dBpT
dB(W/m²)		+30	−10	+25.8	+85.8	+145.8	−25.8	+34.2	+94.2	+96.2
dB(mW/m²)	−30		−40	−4.2	+55.8	+115.8	−55.8	+4.2	+64.2	+66.2
dB(mW/cm²)	+10	+40		+35.8	+95.8	+155.8	−15.8	+44.2	+104.2	+106.2
dB(V/m)	−25.8	+4.2	−35.8		+60	+120	−51.5	+8.5	+68.5	+70.5
dB(mV/m)	−85.8	−55.8	−95.8	−60		+60	−111.5	−51.5	+8.5	+10.5
dB(µV/m)	−145.8	−115.8	−155.8	−120	−60		−171.5	−111.5	−51.5	−49.5
dB(A/m)	+25.8	+55.8	+15.8	+51.5	+111.5	+171.5		+60	+120	+122
dB(mA/m)	−34.2	−4.2	−44.2	−8.5	+51.5	+111.5	−60		+60	+62
dB(µA/m)	−94.2	−64.2	−104.2	−68.5	−8.5	+51.5	−120	−60		+2
dBpT	−96.2	−66.2	−106.2	−70.5	−10.5	+49.5	−122	−62	−2	

V.7.2   Electric and Magnetic Field Strength, (Power) Flux Density Levels

Table V.5 enables simple conversion between electric and magnetic field strength levels and between power flux density levels and magnetic flux density levels. Correct values are achieved under the assumption of free-space propagation of electromagnetic waves under far-field conditions.

For example a *magnetic* field strength level of $-5\,\text{dB}(\mu\text{A/m})$ causes a power flux density level in a field far from the radiation source of

$$-5\,\text{dB}(\mu\text{A/m}) - 94.2 = -99.2\,\text{dB}(\text{W/m}^2)$$

or, expressed in $\text{dB}(\text{mW/m}^2)$,

$$-5\,\text{dB}(\mu\text{A/m}) - 64.2 = -69.2\,\text{dB}(\text{mW/m}^2)$$
$$\widehat{=}$$
$$-99.2\,\text{dB}(\text{W/m}^2) + 30 = -69.2\,\text{dB}(\text{mW/m}^2)$$

or, expressed in $\text{dB}(\text{mW/cm}^2)$

$$-5\,\text{dB}(\mu\text{A/m}) - 104.2 = -109.2\,\text{dB}(\text{mW/cm}^2)$$
$$\widehat{=}$$
$$-99.2\,\text{dB}(\text{W/m}^2) - 10 = -109.2\,\text{dB}(\text{mW/m}^2)$$

and results in an *electric* field strength level of

$$-5\,\text{dB}(\mu\text{A/m}) + 51.5 = 46.5\,\text{dB}(\mu V/\text{m})$$
$$\widehat{=}$$
$$-99.2\,\text{dB}(\text{W/m}^2) + 145.8 = 46.6\,\text{dB}(\mu V/\text{m})$$

with a field wave impedance of the free space at $377\,\Omega$.

References

[1] Ulrich Tietze, Christoph Schenk: Halbleiter-Schaltungstechnik (Circuit Designs Using Semi-Conductors); 12th edition; Springer Verlag 2002; ISBN 978-3-540-42849-7

[2] Ralf Rudersdorfer: Der korrekte Umgang mit Dezibel in übertragungstechnischen Anwendungen (Correct Use of Decibels in Transmission Applications); manuscript of speeches at AFTM Munich 2006, pp. 37–51

[3] Thomas Rühle: Entwurfsmethodik für Funkempfänger – Architekturauswahl und Blockspezifikation unter schwerpunktmäßiger Betrachtung des Direct-Conversion- und des Superheterodynprinzipes (Methodology of Designing Radio Receivers – Architecture Selection and Block Specification with the Main Focus on Direct Conversion and Superheterodyne Designs); dissertation at the TU Dresden 2001

[4] Avago Technologies, editor: AppCAD; www.hp.woodshot.com

[5] Ulrich Graf: Empfänger-Intermodulation – Teil 1 bis Teil 3 (Receiver Intermodulation – Part 1 to Part 3); CQ DL 6/2002, pp. 436–438, CQ DL 7/2002, pp. 504–507, CQ DL 8/2002, pp. 588–591; ISSN 0178-269X

[6] Thomas Moliere: Das Großsignalverhalten von Kurzwellenempfängern (Large-Signal Behaviour of Short-Wave Receivers); CQ DL 8/1973, pp. 450–458; ISSN 0178-269X

[7] Christoph Rauscher: Grundlagen der Spektrumanalyse (Fundamentals of Spectrum Analysis); 2nd edition; Rohde&Schwarz in-house publishing 2004; PW 0002.6629.00

[8] Manfred Thumm, editor: Hoch- und Höchstfrequenz-Halbleiterschaltungen (High and Super High Frequency Semi-Conductor Circuits); manuscript of the Karlsruhe Institute of Technology 10/2008, pp. 1–191

[9] Michael Hiebel: Grundlagen der vektoriellen Netzwerkanalyse (Principles of Vectorial Network Analysis); 2nd edition; Rohde&Schwarz in-house publishing 2007; ISBN 978-3-939837-05-3

[10] International Telecommunication Union (ITU), publisher: Radio Regulations; Edition 2008, Article 2 Nomenclature – Classification of emissions and necessary bandwidths

[11] International Telecommunication Union (ITU), publisher: Radio Regulations; Edition 2008, Article 5 The international Table of Frequency Allocations

Further Reading

Detlef Lechner: Kurzwellenempfänger (Short-Wave Receivers); 2nd edition; Militärverlag der Deutschen Demokratischen Republik 1985

Ferdinand Nibler, editor: Hochfrequenz-Schaltungstechnik (High-Frequency Circuit Designs); 3rd edition; expert Verlag 1998; ISBN 3-8169-1468-3

James Bao-Yen Tsui: Microwave Receivers with Electronic Warfare Applications; reprint edition; Krieger Publishing Company 1992; ISBN 0-89464-724-5

Peter Winterhalder: Intermodulation and noise in receiving systems; News from Rohde&Schwarz 7/1977, pp. 28–31; ISSN 0548–3093

List of Tables

Radio Receiver Technology: Principles, Architectures and Applications, First Edition. Ralf Rudersdorfer.
© 2014 Ralf Rudersdorfer. Published 2014 by John Wiley & Sons, Ltd.

Index